Jan Marius Hofert

Sampling Nested Archimedean Copulas

Jan Marius Hofert

Sampling Nested Archimedean Copulas

with Applications to CDO Pricing

Südwestdeutscher Verlag für Hochschulschriften

Impressum/Imprint (nur für Deutschland/ only for Germany)
Bibliografische Information der Deutschen Nationalbibliothek: Die Deutsche Nationalbibliothek verzeichnet diese Publikation in der Deutschen Nationalbibliografie; detaillierte bibliografische Daten sind im Internet über http://dnb.d-nb.de abrufbar.

Alle in diesem Buch genannten Marken und Produktnamen unterliegen warenzeichen-, marken- oder patentrechtlichem Schutz bzw. sind Warenzeichen oder eingetragene Warenzeichen der jeweiligen Inhaber. Die Wiedergabe von Marken, Produktnamen, Gebrauchsnamen, Handelsnamen, Warenbezeichnungen u.s.w. in diesem Werk berechtigt auch ohne besondere Kennzeichnung nicht zu der Annahme, dass solche Namen im Sinne der Warenzeichen- und Markenschutzgesetzgebung als frei zu betrachten wären und daher von jedermann benutzt werden dürften.

Verlag: Südwestdeutscher Verlag für Hochschulschriften Aktiengesellschaft & Co. KG
Dudweiler Landstr. 99, 66123 Saarbrücken, Deutschland
Telefon +49 681 37 20 271-1, Telefax +49 681 37 20 271-0
Email: info@svh-verlag.de
Zugl.: Ulm, Universität, Dissertation, 2010

Herstellung in Deutschland:
Schaltungsdienst Lange o.H.G., Berlin
Books on Demand GmbH, Norderstedt
Reha GmbH, Saarbrücken
Amazon Distribution GmbH, Leipzig
ISBN: 978-3-8381-1656-3

Imprint (only for USA, GB)
Bibliographic information published by the Deutsche Nationalbibliothek: The Deutsche Nationalbibliothek lists this publication in the Deutsche Nationalbibliografie; detailed bibliographic data are available in the Internet at http://dnb.d-nb.de.

Any brand names and product names mentioned in this book are subject to trademark, brand or patent protection and are trademarks or registered trademarks of their respective holders. The use of brand names, product names, common names, trade names, product descriptions etc. even without a particular marking in this works is in no way to be construed to mean that such names may be regarded as unrestricted in respect of trademark and brand protection legislation and could thus be used by anyone.

Publisher: Südwestdeutscher Verlag für Hochschulschriften Aktiengesellschaft & Co. KG
Dudweiler Landstr. 99, 66123 Saarbrücken, Germany
Phone +49 681 37 20 271-1, Fax +49 681 37 20 271-0
Email: info@svh-verlag.de

Printed in the U.S.A.
Printed in the U.K. by (see last page)
ISBN: 978-3-8381-1656-3

Copyright © 2010 by the author and Südwestdeutscher Verlag für Hochschulschriften Aktiengesellschaft & Co. KG and licensors
All rights reserved. Saarbrücken 2010

To the most important dependence in life,
Love.

Contents

List of Tables		7
List of Figures		9
Abstract		11
1 Copulas		**15**
1.1	Preliminaries	16
1.2	Definition and basic properties of subcopulas and copulas	21
1.3	Sklar's Theorem	26
1.4	Random vectors and copulas	28
1.5	Survival copulas	30
1.6	Symmetries of copulas	31
1.7	Measures of association	32
	1.7.1 Correlation	33
	1.7.2 Measures of dependence	34
	1.7.3 Measures of concordance	35
	1.7.4 Tail dependence	39
1.8	Sampling copulas	40
1.9	Elliptical copulas	42
2 Archimedean and nested Archimedean copulas		**47**
2.1	Archimedean copulas	48
	2.1.1 Archimedean copulas in two dimensions	49
	2.1.2 Archimedean copulas in arbitrary dimensions	50
2.2	Nested Archimedean copulas	54
2.3	Properties of Archimedean and nested Archimedean copulas	58
2.4	Parametric Archimedean families	61

Contents

| | 2.5 | Sampling Archimedean and nested Archimedean copulas | 67 |

3 Numerically sampling nested Archimedean copulas — 71
- 3.1 Numerical inversion of Laplace transforms 72
  - 3.1.1 The Fixed Talbot algorithm 72
  - 3.1.2 The Gaver-Stehfest and Gaver-Wynn-rho algorithm 73
  - 3.1.3 The Laguerre-series algorithm 77
- 3.2 Sampling Archimedean copulas using numerical inversion algorithms . . 78
- 3.3 Experiments and examples 81
  - 3.3.1 A word concerning the implementation 82
  - 3.3.2 Precision and run-time comparison for Clayton's copula 84
  - 3.3.3 A Clayton copula 86
  - 3.3.4 An outer power Clayton copula 87
  - 3.3.5 A nested outer power Clayton copula 89
  - 3.3.6 Nested Gumbel and Clayton copulas 91
  - 3.3.7 An Ali-Mikhail-Haq(Clayton) copula 92
- 3.4 Conclusion 93

4 Directly sampling nested Archimedean copulas — 97
- 4.1 Nested Ali-Mikhail-Haq, Frank, and Joe copulas 98
- 4.2 Nested Archimedean copulas based on generator transformations 103
  - 4.2.1 Exponentially tilted generators and a fast sampling algorithm . . 104
  - 4.2.2 Sampling exponentially tilted stable distributions 106
  - 4.2.3 A general nesting transformation 110
- 4.3 Nested Archimedean copulas based on different Archimedean families . 113
- 4.4 Experiments and examples 118
  - 4.4.1 Joe copulas 118
  - 4.4.2 Nested Frank and Joe copulas 120
  - 4.4.3 Comparison of algorithms for nested Clayton copulas 123
  - 4.4.4 Overall precision and run-time comparison 126
  - 4.4.5 An Ali-Mikhail-Haq(Clayton(20)) copula 127
- 4.5 Conclusion 130

5 CDO pricing with nested Archimedean copulas — 133
- 5.1 Motivation and introduction 134

	5.2	The model	137
		5.2.1 The copulas considered	138
		5.2.2 Default correlations in a nested framework	139
	5.3	Portfolio CDSs and CDOs	140
		5.3.1 The payment streams	141
		5.3.2 The pricing approach	142
	5.4	Calibration to market spreads	145
		5.4.1 Data and setup	146
		5.4.2 Calibration of the model	146
		5.4.3 Results of the calibration	149
	5.5	Conclusion	156

A Supplementary material — 157

A.1 Conditional distribution functions 157
A.2 Marshall-Olkin copulas 158
A.3 Laplace-Stieltjes transforms 160

B Supplementary CDO pricing results — 167

Bibliography — 177

Index — 187

Contents

List of Tables

2.1	One-parameter Archimedean generators with explicit inverse Laplace-Stieltjes transforms.	62
2.2	Kendall's tau and tail-dependence coefficients for commonly used one-parameter Archimedean generators.	65
2.3	Nesting conditions and inverse Laplace-Stieltjes transforms corresponding to the inner generator $\psi_{01}(\,\cdot\,;V_0)$.	65
3.1	Precision and run times in seconds for 1 000 evaluations of $\bar{F}(x)$ for all $x \in P$. .	85
3.2	Precision of the direct approach, the conditional distribution method, and the Fixed Talbot algorithm for generating 100 000 vectors of random variates from a trivariate Clayton copula.	88
3.3	Run times in seconds for the direct approach, the conditional distribution method, and the Fixed Talbot algorithm for generating 100 000 vectors of random variates from a Clayton copula.	88
4.1	Mean run times in seconds for sampling bivariate Joe copulas with two different algorithms for finding quantiles.	119
4.2	Precision and mean run times in seconds for fully-nested trivariate Frank copulas based on 100 000 vectors of random variates.	121
4.3	Precision and mean run times in seconds for fully-nested trivariate Joe copulas based on 100 000 vectors of random variates.	122
4.4	Run times in seconds for generating 100 000 vectors of random variates from fully-nested trivariate Clayton copulas with different parameters.	125
4.5	Precision and mean run times in seconds for drawing 100 000 vectors of random variates from different nested Archimedean copulas.	127
5.1	Results as at 2007-06-12 based on 500 000 runs and $T = 5$.	151
5.2	Results as at 2007-06-12 based on 500 000 runs and $T = 10$.	152

List of Tables

5.3	Average calibration results based on 500 000 runs.	153
5.4	Confidence intervals for the CDO upfront payment and spreads for the outer power Clayton copula fitted to the data as at 2007-06-12 based on 500 000 runs.	154
B.1	Results as at 2007-06-14 based on 500 000 runs and $T = 5$.	168
B.2	Results as at 2007-06-14 based on 500 000 runs and $T = 10$.	169
B.3	Results as at 2007-06-19 based on 500 000 runs and $T = 5$.	170
B.4	Results as at 2007-06-19 based on 500 000 runs and $T = 10$.	171
B.5	Results as at 2007-06-21 based on 500 000 runs and $T = 5$.	172
B.6	Results as at 2007-06-21 based on 500 000 runs and $T = 10$.	173
B.7	Results as at 2007-06-26 based on 500 000 runs and $T = 5$.	174
B.8	Results as at 2007-06-26 based on 500 000 runs and $T = 10$.	175

List of Figures

2.1 Tree structure of a trivariate fully-nested Archimedean copula. 55
2.2 Tree structure of a d-dimensional partially-nested Archimedean copula with $\sum_{s=1}^{S} d_s = d$. 56
3.1 Approximation $\tilde{F}$ of F (left). A scatter plot matrix for the outer power Clayton copula with generator $(1 + t^{2/3})^{-2}$ (right). 89
3.2 Approximation $\tilde{F}_0$ of F_0 (left). A scatter plot matrix for the nested outer power Clayton copula with generators $1/(1 + t^{1/\beta_i})$ and $\beta_i \in \{1.1, 1.5\}$ (right). 90
3.3 F corresponding to $S(1/\vartheta, 1, \cos^\vartheta(\pi/(2\vartheta)), \mathbb{1}_{\{\vartheta=1\}}; 1)$ (left). F_{01} corresponding to $\psi_{01}(t; V_0) = \exp(-V_0((1 + t)^\alpha - 1))$ (right). 92
3.4 Approximations $\tilde{F}_{01}$ of F_{01} (left). A scatter plot matrix for the A(C) copula with generators $0.2/(\exp(t) - 0.8)$ and $(1 + t)^{-1/2}$ (right). 93
4.1 1 000 vectors of random variates of a trivariate fully-nested Frank (left) and Joe (right) copula. The parameters involved are chosen such that Kendall's tau of 0.2 and 0.5 are matched. 123
4.2 Tree structure of a four-dimensional fully-nested Archimedean copula. 128
4.3 1 000 vectors of random variates following a trivariate fully-nested Clayton copula with $\tau(\tau) = 0.2(0.5)$ (left) and a four-dimensional fully-nested A(C(20)) copula with $\tau(\tau(\tau)) = 0.2(0.4(0.7))$ (right). 130
5.1 A CDO depicted at time t, after the first premium payment was made. So far, no defaults occurred. 136
5.2 A CDO depicted at time t, after two assets are defaulted. Default payments took place and the premium payments are reduced accordingly. 136
5.3 Default correlations for the fitted copulas and $T = 5$ as at 2007-06-12. 154

List of Figures

5.4 Progression of the CDO calibration algorithm for the outer power Clayton copula (2007-06-12, $T = 10$). The crosses and circles indicate whether the upfront criterion $D_1 \leq \varepsilon_{\text{CDO}}$ is met or not, respectively. The error D_2, illustrated with different shades of gray, is optimized over all marked points in the $\vartheta_0\vartheta_1$-plane satisfying $D_1 \leq \varepsilon_{\text{CDO}}$. The square and the diamond indicate the minimizing arguments for the Archimedean and nested Archimedean copula, respectively. . . . 155

Abstract

The goal of this dissertation is to explore nested Archimedean copulas. In particular, efficient sampling algorithms, especially suited for large dimensions, are presented. As an application, a pricing model for collateralized debt obligations ("CDOs") is developed.

Copulas are distribution functions with standard uniform univariate margins. It is often criticized that most of the copula theory is developed for two dimensions, although from the practitioner's point of view the studied random vectors usually live in higher dimensions. One particular parametric class of copulas which extends to any multivariate case is the class of elliptical copulas. However, elliptical copulas are radially symmetric which is considered to be a drawback, e.g., for modeling some of the observed effects underlying financial data. A class of copulas that naturally extends to the multivariate case and does not show these deficiencies consists of Archimedean copulas. Archimedean copulas are explicit and can be expressed in terms of a one-dimensional function called the generator of the Archimedean copula. However, these copulas are permutation symmetric in their arguments. To circumvent the rather strong assumption of symmetry in large dimensions, a new class of copulas has recently been suggested, the class of nested Archimedean copulas. Since members of this flexible class are built by nesting Archimedean copulas at different levels, nested Archimedean copulas allow one to precisely capture hierarchical structures often inherent in practical applications.

The construction of parametric copula families is still one of the most active fields of copula research. Such a construction is often accompanied by a sampling algorithm which may help us to understand the induced dependence structure. Sampling algorithms for copulas are also indispensable for the evaluation of copulas that are not explicit, e.g., for elliptical copulas such as the Gaussian or t copula. Moreover, from the practitioner's point of view, fast sampling algorithms are required for simulation studies involving multivariate dependencies, e.g., for pricing complicated financial contracts.

For sampling Archimedean copulas there are at least three known approaches: the conditional distribution method, a method based on a certain transformation, and an algorithm based on Laplace-Stieltjes transforms. The first two are rarely applicable for efficiently sampling

Abstract

Archimedean copulas in large dimensions as they require high-order derivatives of the Archimedean generator to be known. The derivatives involved get even more complicated for nested Archimedean copulas. The conditional distribution method is therefore practically not applicable for sampling nested Archimedean copulas. The transformation method is not even known to extend to that case. However, the algorithm based on Laplace-Stieltjes transforms is efficient for both Archimedean and nested Archimedean copulas, particularly in large dimensions, assuming the Laplace-Stieltjes inverses involved are easy to sample. Finding efficient sampling strategies for these distribution functions on the positive real line is the main goal of this dissertation. As a general-purpose approach, we first investigate several algorithms for numerically inverting Laplace transforms to access these distributions. These numerical approaches seem especially appealing for sampling Archimedean copulas. We also discuss their applicability for sampling nested Archimedean copulas. Our findings are given in Chapter 3.

Note that despite being explicit in their general form, there are not many specific parametric Archimedean generators known that lead to explicit Archimedean copulas, due to the fact that both the generator and its inverse need to be given explicitly. There is currently only one parametric generator known to lead to a straightforward and efficient sampling algorithm for the corresponding family of nested Archimedean copulas. Our goal is therefore to develop fast sampling algorithms for both Archimedean and nested Archimedean copulas constructed with commonly used generators. On our way through the Archimedean world, we obtain general results concerning the construction and sampling of nested Archimedean copulas. One result shows how an Archimedean generator may be transformed to obtain a parametric generator that leads to a nested Archimedean copula which can be sampled. Our findings lead to several byproducts, including an efficient sampling algorithm for exponentially tilted distributions. Exponentially tilted stable distributions appear in the theory and applications of Lévy processes and several approximative algorithms have recently been suggested. In contrary, the algorithm we propose for sampling these distributions is exact. Further results include sampling algorithms for nested Archimedean copulas when the underlying distributions are discrete, the construction of nested Archimedean copulas based on generators belonging to different Archimedean families, and properties about the induced dependence structures. These results are presented in Chapter 4.

CDOs are asset-backed securities which provide an insurance against defaults in a portfolio of assets affected by credit risk. The policyholder is obliged to make regular premium payments to the insurer, whereas the insurer compensates for losses in the portfolio. The portfolio is subdivided into different risk classes, called tranches, which are affected by losses according to their seniority. The goal is to find a premium such that the contract is fair for both parties. Being exposed to the same macro-economic factors, the underlying assets are usually dependent. Therefore, the copula concept emerges. Due to their desirable properties, Archimedean copulas have recently been considered for pricing CDOs. Usually, the portfolio-loss distribution is not known explicitly and the fair spread of a CDO tranche is in general obtained by means of a Monte Carlo simulation. As portfolios are typically high-dimensional, the symmetry of Archimedean copulas is considered to be a drawback. Further, components of such portfolios can typically be assigned to one of several sectors which are affected by different macro-economic factors and therefore behave differently. This is only one application where nested Archimedean copulas arise naturally. We extend a copula-based CDO pricing model to allow for nested Archimedean copulas which precisely capture the structure of the underlying portfolio. Our first goal is to show that such a hierarchical model captures market spreads much more accurately than a non-sectorial, symmetric model. For this purpose, we develop a fast calibration algorithm for CDOs based on nested Archimedean copulas. Although we only consider two-parameter copulas, the resulting pricing error is significantly reduced in comparison to the one-parameter symmetric case, but also in comparison to the Gauss copula which we include in our investigation as the standard market model. Our studies strongly suggest the use of hierarchical models for pricing CDOs. This part of this work is developed in collaboration with Matthias Scherer and presented in Chapter 5.

Abstract

1 Copulas

> "Knowing the word 'copula' as a grammatical term for a word or expression that links a subject and predicate, I felt that this would make an appropriate name for a function that links a multidimensional distribution to its one-dimensional margins and used it as such."
>
> Sklar (1996)

In short, a d-dimensional copula is a d-dimensional distribution function with standard uniform univariate margins. The investigation of standardized distribution functions dates back to the work of Hoeffding (1940), who obtains many results known today for standardized distribution functions on the square with sides $[-1/2, 1/2]$. Féron (1956) considers standardized distribution functions on the three-dimensional unit cube. Sklar (1959) introduces the term "copula", Latin for "link", and proves an important result, known as Sklar's Theorem. In that result, he considers copulas in their general form and establishes a link between multivariate distribution functions and their one-dimensional margins.

As Schweizer (1991) points out in his historical overview, in the beginning of copula theory after the work of Sklar (1959), virtually all results were obtained in the context of probabilistic metric spaces. With the work of Schweizer and Wolff (1981), who investigate criteria for measures of dependence, copulas turn out to provide a simple tool to analyze the dependence between two or more random variables. These authors also show that copulas are invariant under strictly increasing transformations. In combination with Sklar's Theorem, this implies that copulas precisely capture the information about the dependence structure between random variables. By the middle of the nineties of the last century, copulas enter the world of financial and insurance mathematics and are used to model various kinds of dependencies. Since then, this young and active field of research has developed quite fast, see, e.g., Genest et al. (2009). The subprime mortgage crisis shows that adequate modeling of dependence structures is crucial,

1 Copulas

e.g., for pricing financial products with dependent underlying components.

The two main reasons for the popularity of copulas, see, e.g., Fisher (1997), are that copulas can be used to describe and investigate dependence between random variables; further, copulas can serve as building blocks for multivariate distribution functions, often in the context of simulation algorithms.

In the following sections we give an introduction to copulas and related concepts of measuring dependence between random variables. The theory of copulas is a rather young mathematical subject and some proofs are scattered in various research papers, not always presented in a concise mathematical way, and often only provided for the bivariate case. We therefore give a self-contained introduction to the topic, including proofs for most results. We mainly focus on the multivariate case which is especially desirable in applications. As an exception, measures of association are presented only for the bivariate case, as they are much more common than their possible multivariate extensions. For the sake of readability, some background results are given in the appendix. It is clear that such an introduction can by no means be exhaustive. We therefore only present the results needed in this work and refer the interested reader to the more detailed literature, including the textbooks of Schweizer and Sklar (1983), Joe (1997), McNeil et al. (2005), and Nelsen (2007).

1.1 Preliminaries

In this section, we present definitions and basic properties of the concepts needed later on. By an *increasing* function we mean a non-decreasing function. Unless otherwise stated, all inequalities, intervals, and statements involving vectors are understood to hold componentwise.

Definition 1.1.1
Let $\emptyset \neq S_j \subseteq \bar{\mathbb{R}} := \mathbb{R} \cup \{-\infty, \infty\}$, $j \in \{1, \ldots, d\}$, and assume that $s_{j,l} := \min S_j$ and $s_{j,u} := \max S_j$, $j \in \{1, \ldots, d\}$, exist in the improper sense, i.e., possibly $s_{j,l} = -\infty$ and $s_{j,u} = \infty$. Let $H : \prod_{j=1}^{d} S_j \to \mathbb{R}$, where $H(\boldsymbol{x})$ is set to the corresponding limit, assumed to exist, if at least one of the components of $\boldsymbol{x}$ is $-\infty$ or ∞. For $j \in \{1, \ldots, d\}$ define

$$\Delta_{(a_j, b_j]}^0 H(\boldsymbol{x}) := H(\boldsymbol{x}),$$
$$\Delta_{(a_j, b_j]} H(\boldsymbol{x}) := H(x_1, \ldots, x_{j-1}, b_j, x_{j+1}, \ldots, x_d) - H(x_1, \ldots, x_{j-1}, a_j, x_{j+1}, \ldots, x_d),$$
$$\Delta_{(a_j, b_j]}^k H(\boldsymbol{x}) := \Delta_{(a_j, b_j]}(\Delta_{(a_j, b_j]}^{k-1} H(\boldsymbol{x})), \ k \in \mathbb{N},$$

where all arguments of H are assumed to lie in $\prod_{j=1}^{d} S_j$. The H-*volume* $\Delta_{(\boldsymbol{a}, \boldsymbol{b}]} H$ is defined by $\Delta_{(\boldsymbol{a}, \boldsymbol{b}]} H := \Delta_{(a_d, b_d]} \ldots \Delta_{(a_1, b_1]} H$ for all $\boldsymbol{a}, \boldsymbol{b} \in \prod_{j=1}^{d} S_j : \boldsymbol{a} \leq \boldsymbol{b}$. H is called *grounded* if $x_j = s_{j,l}$ for

1.1 Preliminaries

at least one $j \in \{1,\ldots,d\}$ implies $H(\boldsymbol{x}) = 0$. The j-th *margin* F_j, $j \in \{1,\ldots,d\}$, of H is defined as $F_j(x_j) := H(s_{1,u}, \ldots, s_{j-1,u}, x_j, s_{j+1,u}, \ldots, s_{d,u})$, $x_j \in S_j$. Further, H is called d-*increasing* if $\Delta_{(\boldsymbol{a},\boldsymbol{b}]} H \geq 0$ for all intervals $(\boldsymbol{a}, \boldsymbol{b}]$ with $\boldsymbol{a}, \boldsymbol{b} \in \prod_{j=1}^d S_j : \boldsymbol{a} \leq \boldsymbol{b}$.

Note that the H-volume can be written as

$$\Delta_{(\boldsymbol{a},\boldsymbol{b}]} H = \sum_{i_d=1}^{2} \cdots \sum_{i_1=1}^{2} (-1)^{i_1+\cdots+i_d} H(x_{1i_1}, \ldots, x_{d i_d}). \tag{1.1}$$

where $x_{j1} = a_j$, $x_{j2} = b_j$ for all $j \in \{1,\ldots,d\}$. It follows that $\Delta_{(\boldsymbol{a},\boldsymbol{b}]} H = \Delta_{(a_{j_d}, b_{j_d}]} \cdots \Delta_{(a_{j_1}, b_{j_1}]} H$ for all permutations $(j_1, \ldots, j_d)$ of $\{1,\ldots,d\}$. Although the following proposition simply combines the fact that H is grounded and d-increasing, it leads to several results about functions sharing these properties.

Proposition 1.1.2
Let $\emptyset \neq S_j \subseteq \bar{\mathbb{R}}$, $j \in \{1,\ldots,d\}$, and assume that $s_{j,l} := \min S_j$, $j \in \{1,\ldots,d\}$, exist in the improper sense. Let $H : \prod_{j=1}^d S_j \to \mathbb{R}$ be grounded and d-increasing, and for $k \in \{1,\ldots,d\}$, let $j_1, \ldots, j_k \in \{1,\ldots,d\}$ with $j_i \neq j_l$ for all $i \neq l$. Then

$$\Delta_{(a_{j_k}, b_{j_k}]} \cdots \Delta_{(a_{j_1}, b_{j_1}]} H(\boldsymbol{x}) \geq 0$$

for all $\boldsymbol{x} \in \prod_{j=1}^d S_j$ and $a_{j_l}, b_{j_l} \in S_{j_l} : a_{j_l} \leq b_{j_l}$, $l \in \{1,\ldots,k\}$.

Proof
Let $j_1, \ldots, j_k$, a_{j_l}, b_{j_l}, $l \in \{1,\ldots,k\}$, and $\boldsymbol{x}$ be as in the claim. Extend a_{j_l}, b_{j_l}, $l \in \{1,\ldots,k\}$, to vectors $\boldsymbol{a}', \boldsymbol{b}' \in \prod_{j=1}^d S_j$ with components

$$a'_j = \begin{cases} a_j, & j \in \{j_1,\ldots,j_k\}, \\ s_{j,l}, & \text{otherwise,} \end{cases} \quad b'_j = \begin{cases} b_j, & j \in \{j_1,\ldots,j_k\}, \\ x_j, & \text{otherwise.} \end{cases}$$

As $\Delta_{(a_{j_k}, b_{j_k}]} \cdots \Delta_{(a_{j_1}, b_{j_1}]} H(\boldsymbol{x})$ is invariant under the order of taking differences and since H is grounded, one has $\Delta_{(a_{j_k}, b_{j_k}]} \cdots \Delta_{(a_{j_1}, b_{j_1}]} H(\boldsymbol{x}) = \Delta_{(\boldsymbol{a}', \boldsymbol{b}']} H$, which is non-negative since H is d-increasing. □

As a direct consequence of Proposition 1.1.2, one obtains the following result. It implies that continuity of the margins F_j, $j \in \{1,\ldots,d\}$, leads to continuity of the function H.

Theorem 1.1.3
Let $\emptyset \neq S_j \subseteq \bar{\mathbb{R}}$, $j \in \{1,\ldots,d\}$, and assume that $\min S_j$ and $\max S_j$, $j \in \{1,\ldots,d\}$, exist in

1 Copulas

the improper sense. Let $H : \prod_{j=1}^{d} S_j \to \mathbb{R}$ be grounded and d-increasing, with margins F_j, $j \in \{1, \ldots, d\}$. Then, for all $\boldsymbol{a}, \boldsymbol{b} \in \prod_{j=1}^{d} S_j$,

$$|H(\boldsymbol{b}) - H(\boldsymbol{a})| \leq \sum_{j=1}^{d} |F_j(b_j) - F_j(a_j)|.$$

Proof

The triangle inequality implies $|H(\boldsymbol{b}) - H(\boldsymbol{a})| \leq \sum_{j=1}^{d} |H(b_1, \ldots, b_{j-1}, b_j, a_{j+1}, \ldots, a_d) - H(b_1, \ldots, b_{j-1}, a_j, a_{j+1}, \ldots, a_d)|$. Now consider the j-th summand. If $a_j \leq b_j$, the case $k = 1$ in Proposition 1.1.2 implies that $0 \leq H(b_1, \ldots, b_{j-1}, b_j, a_{j+1}, \ldots, a_d) - H(b_1, \ldots, b_{j-1}, a_j, a_{j+1}, \ldots, a_d)$. Since $\min S_j$ and $\max S_j$ exist, and by the case $k = 2$ in Proposition 1.1.2, this is less than or equal to $F_j(b_j) - F_j(a_j)$. Thus, we obtain $|H(b_1, \ldots, b_{j-1}, b_j, a_{j+1}, \ldots, a_d) - H(b_1, \ldots, b_{j-1}, a_j, a_{j+1}, \ldots, a_d)| \leq |F_j(b_j) - F_j(a_j)|$. The latter inequality also holds for $a_j > b_j$ and therefore completes the proof. □

Assuming that the partial derivatives of $H : \prod_{j=1}^{d} S_j \to \mathbb{R}$ exist, define for $\boldsymbol{a} \in \prod_{j=1}^{d} S_j$

$$D_{j_k \ldots j_1} H(\boldsymbol{a}) := \left. \frac{\partial^k}{\partial x_{j_k} \ldots \partial x_{j_1}} H(\boldsymbol{x}) \right|_{\boldsymbol{x} = \boldsymbol{a}},$$

where $j_l \in \{1, \ldots, d\}$, $l \in \{1, \ldots, k\}$, and $j_i \neq j_l$ for all $i \neq l$. The following result shows that d-increasing functions living on a reasonable domain allow for first-order partial derivatives almost everywhere with respect to the Lebesgue measure.

Theorem 1.1.4

Let $S_j = [s_{j,l}, s_{j,u}] \in \bar{\mathbb{R}} : s_{j,l} < s_{j,u}$, $j \in \{1, \ldots, d\}$. Further, let $H : \prod_{j=1}^{d} S_j \to \mathbb{R}$ be d-increasing. Then, for $j \in \{1, \ldots, d\}$ and $x_k \in S_k$, $k \in \{1, \ldots, d\} \setminus \{j\}$, the first-order partial derivative $x_j \mapsto D_j H(\boldsymbol{x})$ exists almost everywhere on S_j with respect to the Lebesgue measure and for such x_j,

$$\Delta_{(a_d, b_d]} \ldots \Delta_{(a_{j+1}, b_{j+1}]} \Delta_{(a_{j-1}, b_{j-1}]} \ldots \Delta_{(a_1, b_1]} D_j H(\boldsymbol{x}) \geq 0$$

for all $a_k, b_k \in S_k : a_k \leq b_k$, $k \in \{1, \ldots, d\} \setminus \{j\}$. In particular, $x_j \mapsto D_j H(\boldsymbol{x}^*)$ is non-negative and $x_k \mapsto D_j H(\boldsymbol{x})$, $k \in \{1, \ldots, d\} \setminus \{j\}$ is increasing.

Proof

By Proposition 1.1.2,

$$\Delta_{(a_j, b_j]} \Delta_{(a_d, b_d]} \ldots \Delta_{(a_{j+1}, b_{j+1}]} \Delta_{(a_{j-1}, b_{j-1}]} \ldots \Delta_{(a_1, b_1]} H(\boldsymbol{x}) \geq 0$$

which implies that $x_j \mapsto \Delta_{(a_d,b_d]} \ldots \Delta_{(a_{j+1},b_{j+1}]}\Delta_{(a_{j-1},b_{j-1}]} \ldots \Delta_{(a_1,b_1]}H(\boldsymbol{x})$ is increasing on S_j and therefore differentiable almost everywhere on S_j with respect to the Lebesgue measure, see Folland (1999, p. 101). Thus, the first part of the claim follows by interchanging the operator D_j and the difference operators. The remaining part of the claim follows directly. □

In the sequel, we work on a probability space $(\Omega, \mathcal{F}, \mathbb{P})$ with sample space Ω, σ-algebra $\mathcal{F}$, and probability measure $\mathbb{P}$, on which all considered random variables are assumed to be defined. If $\boldsymbol{X} : \Omega \to \mathbb{R}^d$ is a d-dimensional random vector, its *probability distribution function*, or simply *distribution function*, $H : \bar{\mathbb{R}}^d \to I := [0,1]$ is defined by $H(\boldsymbol{x}) := \mathbb{P}(\boldsymbol{X} \leq \boldsymbol{x})$, $\boldsymbol{x} \in \bar{\mathbb{R}}^d$, where $H(\boldsymbol{x})$ is set to the corresponding limit if at least one of the components of $\boldsymbol{x}$ is $-\infty$ or ∞. A distribution function H is easily shown to be grounded. Further, H is equal to its j-th *margin* $F_j(x_j)$ if $x_k = \infty$ for all $k \in \{1, \ldots, d\}\setminus\{j\}$. Moreover, H trivially assigns non-negative mass to any non-empty d-dimensional interval, so H is d-increasing. Therefore, the preceding results apply to distribution functions. Moreover, H is right-continuous, i.e., $\lim_{\boldsymbol{h}\downarrow \boldsymbol{0}} H(\boldsymbol{x}+\boldsymbol{h}) = H(\boldsymbol{x})$, for all $\boldsymbol{x} \in \mathbb{R}^d$. The *survival function* $\bar{H} : \bar{\mathbb{R}}^d \to I$ of H is defined by $\bar{H}(\boldsymbol{x}) := \mathbb{P}(\boldsymbol{X} > \boldsymbol{x})$, $\boldsymbol{x} \in \bar{\mathbb{R}}^d$, where $H(\boldsymbol{x})$ is again set to the corresponding limit if at least one of the components of $\boldsymbol{x}$ is $-\infty$ or ∞.

The following definition generalizes the notion of an inverse to increasing functions.

Definition 1.1.5
For an increasing function $T : \bar{\mathbb{R}} \to \mathbb{R}$ with $T(-\infty) := \lim_{x \to -\infty} T(x)$ and $T(\infty) := \lim_{x \to \infty} T(x)$, the *generalized inverse* $T^- : \mathbb{R} \to \bar{\mathbb{R}}$ of T is defined by

$$T^-(y) := \inf\{x \in \mathbb{R} : T(x) \geq y\},$$

with the convention that $\inf \emptyset := \infty$. If $T : \bar{\mathbb{R}} \to I$ is a distribution function, $T^- : I \to \bar{\mathbb{R}}$ is also called the *quantile function* of T.

If T is continuous and strictly increasing, T^- coincides with T^{-1}, the ordinary inverse of T. Note that $T^-(y) = -\infty$ if and only if $T(x) \geq y$ for all $x \in \mathbb{R}$. Similarly, $T^-(y) = \infty$ if and only if $T(x) < y$ for all $x \in \mathbb{R}$. When working with generalized inverses, the following result is often useful, where $\operatorname{ran} T$ denotes the range of T.

Proposition 1.1.6
Let $T : \bar{\mathbb{R}} \to \mathbb{R}$ be increasing with $T(-\infty) := \lim_{x \to -\infty} T(x)$ and $T(\infty) := \lim_{x \to \infty} T(x)$. Further, let $x, y \in \mathbb{R}$. Then,

(1) T^- is increasing. If $T^-(y) < \infty$, then T^- is left-continuous in y, and if $T^-(y) > -\infty$, then T^- admits a limit from the right at y.

1 Copulas

(2) T is continuous if and only if T^- is strictly increasing on $\operatorname{ran} T$. T is strictly increasing if and only if T^- is continuous on $\operatorname{ran} T$.

(3) $T^-(T(x)) \leq x$. If T is strictly increasing then $T^-(T(x)) = x$.

(4) If T is right-continuous, then $T^-(y) < \infty$ implies $T(T^-(y)) \geq y$, and $y \in \operatorname{ran} T$ implies $T(T^-(y)) = y$.

(5) $T(x) \geq y$ implies $x \geq T^-(y)$. If T is right-continuous, then $x \geq T^-(y)$ implies $T(x) \geq y$.

(6) If T_1 and T_2 are right-continuous transformations with properties as T, then $(T_1 \circ T_2)^- = T_2^- \circ T_1^-$.

Proof

(1) T^- is increasing since $\{x \in \mathbb{R} : T(x) \geq y_2\} \subseteq \{x \in \mathbb{R} : T(x) \geq y_1\}$ for all $y_1, y_2 \in \mathbb{R} : y_1 < y_2$. To show left-continuity in $y =: y_0 \in \mathbb{R} : T^-(y_0) < \infty$, let $(y_n)_{n \in \mathbb{N}} \subseteq \mathbb{R} : y_n \uparrow y_0$. Then $x_n := T^-(y_n) \leq x_0 := T^-(y_0)$, thus $x_n \nearrow x \leq x_0$ for $n \to \infty$ for some $x \in \mathbb{R}$. By definition of T^-, $T(x_n - \varepsilon) < y_n \leq T(x_n + \varepsilon)$ for all $\varepsilon > 0$ and $n \in \mathbb{N}_0$. If $x < x_0$, then $\varepsilon := (x_0 - x)/2$ implies $y_n \leq T(x_n + \varepsilon) \leq T(x_0 - \varepsilon) < y_0$, thus $y_0 = \lim_{n \to \infty} y_n \leq T(x_0 - \varepsilon) < y_0$, a contradiction. To show that T^- admits a limit from the right at y, let $y_n \downarrow y \in \mathbb{R} : T^-(y) > -\infty$ and note that $(T^-(y_n))_{n \in \mathbb{N}}$ is decreasing and bounded from below by $T^-(y)$.

(2) For the first statement, let $T(x-)$ and $T(x+)$ denote the limits from the left and right, respectively. T has a discontinuity at x if and only if $y_1 := T(x-) < T(x+) =: y_2$, which in turn happens if and only if $T(x - \varepsilon) \leq y_1 < y_2 \leq T(x + \varepsilon)$ for all $\varepsilon > 0$. By definition of T^-, the latter condition is equivalent to $x - \varepsilon \leq T^-(y_1) \leq T^-(y_2) \leq x + \varepsilon$ for all $\varepsilon > 0$ and therefore to $T^-(y_1) = T^-(y_2) = x$, which is equivalent to T^- being not strictly increasing on $\operatorname{ran} T$. For the second statement, note that T is not strictly increasing if and only if there exist $x_1, x_2 \in \mathbb{R} : x_1 < x_2$ and $T(x_1) = T(x_2) =: y$. By definition of T^-, $T(x_2) = y < y + \varepsilon$ implies $T^-(y + \varepsilon) \geq x_2$ for all $\varepsilon > 0$, thus $T^-(y+) \geq x_2$. This implies that $T(x_1) = T(x_2) = y$ if and only if $T^-(y) \leq x_1 < x_2 \leq T^-(y+)$, which is equivalent to T^- being discontinuous at $y \in \operatorname{ran} T$.

(3) The first part follows by definition of T^-. For the second part, note that T being strictly increasing implies that there is no $z < x$ with $T(z) \geq T(x)$, thus $T^-(T(x)) \geq x$.

(4) For the first part, $T^-(y) < \infty$ implies that $A := \{x \in \mathbb{R} : T(x) \geq y\} \neq \emptyset$; thus, there exists $(x_n)_{n \in \mathbb{N}} \subseteq A$ with $x_n \downarrow \inf A = T^-(y)$ for $n \to \infty$. By right-continuity of T, $T(T^-(y)) \nearrow T(x_n) \geq y$, so $T(T^-(y)) \geq y$. For the second part, proceed similarly. Let

$y \in \operatorname{ran} T$ and consider $A := \{x \in \mathbb{R} : T(x) = y\} \neq \emptyset$. Note that $\inf A = T^-(y)$ and conclude that $T(T^-(y)) \nearrow T(x_n) = y$, thus $T(T^-(y)) = y$.

(5) The first statement follows by definition of T^-. For the second statement, note that $T^-(y) \leq x$ implies $y \leq T(T^-(y)) \leq T(x)$, where $y \leq T(T^-(y))$ follows by right-continuity of T.

(6) Applying (5) to T_1 and T_2 leads to $(T_1 \circ T_2)^-(y) = \inf\{x \in \mathbb{R} : T_1(T_2(x)) \geq y\} = \inf\{x \in \mathbb{R} : x \geq T_2^-(T_1^-(y))\} = T_2^-(T_1^-(y))$.

□

The following proposition allows to transform continuous margins into uniform distributions. Further, its second part is useful for generating random numbers following univariate distributions, known as *inversion method*.

Proposition 1.1.7

Let $X \sim F$.

(1) If F is continuous, then $F(X) \sim \mathrm{U}[0,1]$.

(2) If $U \sim \mathrm{U}[0,1]$, then $F^-(U) \sim F$.

Proof

(1) Since F is continuous, Proposition 1.1.6 (5) implies $\mathbb{P}(F(X) \leq x) = 1 - \mathbb{P}(F(X) \geq x) = 1 - \mathbb{P}(X \geq F^-(x)) = \mathbb{P}(X \leq F^-(x)) = F(F^-(x))$ which equals x for all $x \in (0,1)$ by Proposition 1.1.6 (4).

(2) The claim follows directly by Proposition 1.1.6 (5).

□

With these tools at hand, we are now ready to study copulas.

1.2 Definition and basic properties of subcopulas and copulas

We start with presenting the notion of subcopulas and copulas.

Definition 1.2.1

A d-dimensional *subcopula* is a function $C' : \prod_{j=1}^d S_j \to I$, $\{0,1\} \in S_j \subseteq I$, $j \in \{1, \ldots, d\}$, which

(1) is *grounded*, i.e., $C'(\boldsymbol{u}) = 0$ if $u_j = 0$ for at least one $j \in \{1, \ldots, d\}$;

(2) has *uniform margins*, i.e., $C'(\boldsymbol{u}) = u_j$ for all $u_j \in S_j$ if $u_k = 1$ for all $k \in \{1, \ldots, d\} \backslash \{j\}$;

1 Copulas

(3) is *d-increasing*, i.e., $\Delta_{(a,b]} C' \geq 0$ for all $a, b \in \prod_{j=1}^{d} S_j : a \leq b$.

A d-dimensional *copula* is a d-dimensional subcopula with domain I^d.

By definition, d-dimensional copulas are d-dimensional distribution functions with standard uniform univariate margins. For any $d \geq 3$ and d-dimensional copula C, each k-dimensional margin of C is a k-dimensional copula, $k \in \{2, \ldots, d-1\}$. Although we are mainly interested in copulas, subcopulas turn out to be helpful in the proofs of many statements about copulas.

In addition to the properties provided in Section 1.1, uniform margins additionally lead to the following results. Note that Lipschitz continuity implies absolute continuity.

Proposition 1.2.2

(1) Subcopulas are Lipschitz continuous with Lipschitz constant one.

(2) Subcopulas are uniformly equicontinuous on their domain.

(3) For subcopulas with domain as in Theorem 1.1.4, the partial derivative $D_j C'$, which exists almost everywhere, is bounded from above by one and $D_j C'(u)$ is integrable in u_j with respect to the Lebesgue measure for $j \in \{1, \ldots, d\}$.

Proof

By using uniform margins in Theorem 1.1.3, (1), (2), and the first statement of (3) follow immediately. For the second statement of (3), the Fundamental Theorem of Calculus for Lebesgue Integrals, see, e.g., Folland (1999, p. 106), applies since subcopulas with domain as stated are absolutely continuous in each component. $\square$

In the sequel, we often need to extend a subcopula to a copula. The following lemma gives such a construction via multilinear interpolation. The resulting copula is called *standard extension copula*.

Lemma 1.2.3

Let C' be a subcopula with domain $\prod_{j=1}^{d} S_j$. Then there exists a copula C such that $C(u) = C'(u)$ for all $u \in \prod_{j=1}^{d} S_j$.

Proof

By uniform continuity, extend C' to a subcopula C'' with domain $\prod_{j=1}^{d} \bar{S}_j$. If $\prod_{j=1}^{d} \bar{S}_j = I^d$, the proof is established. Otherwise, proceed as follows. For $x \in I^d \setminus \prod_{j=1}^{d} \bar{S}_j$, let $a, b \in \prod_{j=1}^{d} \bar{S}_j$ such that for each $j \in \{1, \ldots, d\}$, $a_j := \max\{s \in \bar{S}_j : s \leq x_j\}$ and $b_j := \min\{s \in \bar{S}_j : s \geq x_j\}$. Note

that a_j and b_j exist since $\bar{S}_j$ is closed. Define the function $C : I^d \to I$ at $\boldsymbol{x}$ via
$$C(\boldsymbol{x}) := \sum_{\boldsymbol{i} \in \{1,2\}^d} C''(\boldsymbol{u_i}) \lambda(\boldsymbol{x}; \boldsymbol{u_i}), \tag{1.2}$$
where $\boldsymbol{i} := (i_1, \ldots, i_d)^T$, $\boldsymbol{u_i} := (u_{1i_1}, \ldots, u_{di_d})^T$ with $u_{j1} := a_j$, $u_{j2} := b_j$ for all $j \in \{1, \ldots, d\}$, and $\lambda(\boldsymbol{x}; \boldsymbol{u_i}) := \prod_{j=1}^d \lambda_j(x_j; u_{ji_j})$ with
$$\lambda_j(x_j; u_{ji_j}) := \begin{cases} (b_j - x_j)/(b_j - a_j), & a_j < b_j, \ u_{ji_j} = a_j, \\ (x_j - a_j)/(b_j - a_j), & a_j < b_j, \ u_{ji_j} = b_j, \\ 1/2, & a_j = b_j, \end{cases}$$

for all $j \in \{1, \ldots, d\}$. By construction $C(\boldsymbol{u}) = C'(\boldsymbol{u})$ for all $\boldsymbol{u} \in \prod_{j=1}^d S_j$. It remains to show that C is a copula. If $x_j = 0$ for at least one $j \in \{1, \ldots, d\}$, then $a_j = 0$. Since C'' is grounded, all summands in (1.2) with $u_{ji_j} = a_j$ disappear. Further, if $u_{ji_j} = b_j > a_j$, then $\lambda_j(x_j; u_{ji_j}) = 0$, so $C(\boldsymbol{x}) = 0$, and if $u_{ji_j} = b_j = a_j$, then $C(\boldsymbol{x}) = 0$ again holds. Thus, C is grounded. Now assume $a_j < b_j$ for all $j \in \{1, \ldots, d\}$ and let $x_k = 1$ for all $k \in \{1, \ldots, d\} \setminus \{j\}$. This implies that $b_k = 1$ and therefore $\lambda_k(a_k; u_{ki_k}) = 0$ for all $k \in \{1, \ldots, d\} \setminus \{j\}$. Moreover, $\lambda_k(b_k; u_{ki_k}) = 1$ for all $k \in \{1, \ldots, d\} \setminus \{j\}$. Thus, $C(1, \ldots, 1, x_j, 1, \ldots, 1) = C''(1, \ldots, 1, b_j, 1, \ldots, 1)(x_j - a_j)/(b_j - a_j) + C''(1, \ldots, 1, a_j, 1, \ldots, 1)(b_j - x_j)/(b_j - a_j) = x_j$ since C'' has uniform margins. The degenerate case where $a_j = b_j$ for at least one $j \in \{1, \ldots, d\}$ reduces to the case where the same reasoning as above can be applied. Thus, C has uniform margins. Now let $[\boldsymbol{a}, \boldsymbol{b}] \subseteq I^d$. The j-th edge $[a_j, b_j]$ of $[\boldsymbol{a}, \boldsymbol{b}]$ is called *good* if $a_j, b_j \in \bar{S}_j$. Otherwise, it is called *bad*. To show that C is d-increasing, proceed by induction on the number of bad edges of $[\boldsymbol{a}, \boldsymbol{b}]$. If there are no bad edges, every vertex of $[\boldsymbol{a}, \boldsymbol{b}]$ is in the domain of C'', thus $\Delta_{(\boldsymbol{a},\boldsymbol{b})} C = \Delta_{(\boldsymbol{a},\boldsymbol{b})} C'' \geq 0$ since C'' is d-increasing. Assume that for some $n \in \{1, \ldots, d-1\}$, $\Delta_{(\boldsymbol{a},\boldsymbol{b})} C \geq 0$ for all $[\boldsymbol{a}, \boldsymbol{b}] \subseteq I^d$ with less than or equal to n bad edges. Assume $[\boldsymbol{a}, \boldsymbol{b}] \subseteq I^d$ has $n+1$ bad edges and that $[a_j, b_j]$ is bad. If $(a_j, b_j) \cap \bar{S}_j = \emptyset$, define $a_{j,l} := \max\{s \in \bar{S}_j : s \leq a_j\}$ and $b_{j,u} := \min\{s \in \bar{S}_j : s \geq b_j\}$. Then $[\boldsymbol{a}, \boldsymbol{b}] \subseteq A := [a_1, b_1] \times \cdots \times [a_{j-1}, b_{j-1}] \times [a_{j,l}, b_{j,u}] \times [a_{j+1}, b_{j+1}] \times \cdots \times [a_d, b_d]$. If $a_j = b_j$, then $\Delta_{(\boldsymbol{a},\boldsymbol{b})} C = 0$, by definition. Otherwise, $a_{j,l} < b_{j,u}$ and $\Delta_{(\boldsymbol{a},\boldsymbol{b})} C = ((b_j - a_j)/(b_{j,u} - a_{j,l})) \Delta_A C \geq 0$, since A has only n bad edges. Now assume $(a_j, b_j) \cap \bar{S}_j \neq \emptyset$. In addition to $a_{j,l}$ and $b_{j,u}$ as before, define $a_{j,u} := \min\{s \in \bar{S}_j : s \geq a_j\}$ and $b_{j,l} := \max\{s \in \bar{S}_j : s \leq b_j\}$. Split $[\boldsymbol{a}, \boldsymbol{b}]$ into the intervals A_1, A_2, and A_3, given by
$$A_1 := [a_1, b_1] \times \cdots \times [a_{j-1}, b_{j-1}] \times [a_j, a_{j,u}] \times [a_{j+1}, b_{j+1}] \times \cdots \times [a_d, b_d],$$
$$A_2 := [a_1, b_1] \times \cdots \times [a_{j-1}, b_{j-1}] \times [a_{j,u}, b_{j,l}] \times [a_{j+1}, b_{j+1}] \times \cdots \times [a_d, b_d],$$
$$A_3 := [a_1, b_1] \times \cdots \times [a_{j-1}, b_{j-1}] \times [b_{j,l}, b_j] \times [a_{j+1}, b_{j+1}] \times \cdots \times [a_d, b_d].$$

1 Copulas

Since A_2 has only n bad edges, $\Delta_{A_2} C \geq 0$. To see that $\Delta_{A_1} C \geq 0$ and $\Delta_{A_3} C \geq 0$, enlarge A_1 to A'_1 and A_3 to A'_3, with

$$A'_1 := [a_1, b_1] \times \cdots \times [a_{j-1}, b_{j-1}] \times [a_{j,l}, a_{j,u}] \times [a_{j+1}, b_{j+1}] \times \cdots \times [a_d, b_d],$$
$$A'_3 := [a_1, b_1] \times \cdots \times [a_{j-1}, b_{j-1}] \times [b_{j,l}, b_{j,u}] \times [a_{j+1}, b_{j+1}] \times \cdots \times [a_d, b_d]$$

and proceed as for the set A above. Thus, $\Delta_{(a,b]} C = \Delta_{A_1} C + \Delta_{A_2} C + \Delta_{A_3} C \geq 0$. □

The following theorem shows that copulas are bounded by the so-called *lower* and *upper Fréchet bounds* W and M, respectively. This result is attributed to Fréchet (1951), partly appearing in Fréchet (1935) already.

Theorem 1.2.4 (Fréchet bounds)
For $d \geq 2$, define $W(\boldsymbol{u}) := \max\{\sum_{j=1}^{d} u_j - d + 1, 0\}$ and $M(\boldsymbol{u}) := \min_{1 \leq j \leq d}\{u_j\}$, $\boldsymbol{u} \in I^d$. Then

$$W(\boldsymbol{u}) \leq C(\boldsymbol{u}) \leq M(\boldsymbol{u}), \quad \boldsymbol{u} \in I^d.$$

For $d = 2$, W is a copula. For $d \geq 3$ and any $\boldsymbol{u} \in I^d$ there exists a copula C such that $C(\boldsymbol{u}) = W(\boldsymbol{u})$. However, W is not a copula. Further, M is a copula for all $d \geq 2$.

Proof
By Proposition 1.1.2, $C(\boldsymbol{u}) \leq C(1, \ldots, 1, u_j, 1, \ldots, 1) = u_j$ for all $j \in \{1, \ldots, d\}$. Therefore, $C(\boldsymbol{u}) \leq M(\boldsymbol{u})$. Note that M is the distribution function of the random vector $(U, \ldots, U)^T$, $U \sim U[0,1]$. Thus, M is a copula. Let $\mathbf{1} := (1, \ldots, 1)^T \in \mathbb{R}^d$. By Theorem 1.1.3, $1 - C(\boldsymbol{u}) = |C(\mathbf{1}) - C(\boldsymbol{u})| \leq \sum_{j=1}^{d} |1 - u_j| = d - \sum_{j=1}^{d} u_j$, $\boldsymbol{u} \in I^d$, and by definition, $C(\boldsymbol{u}) \geq 0$ for all $u \in I^d$. Thus, $C(\boldsymbol{u}) \geq W(\boldsymbol{u})$ for all $u \in I^d$. For $d = 2$, W is the distribution function of the random vector $(U, 1-U)^T$, $U \sim U[0,1]$. Therefore, by definition, W is a copula. If $d \geq 3$, it follows from Identity (1.1) that $\Delta_{(1/2, 1]} W = \max\{1 + \cdots + 1 - d + 1, 0\} - d \max\{1/2 + 1 + \cdots + 1 - d + 1, 0\} + \binom{d}{2} \max\{1/2 + 1/2 + 1 + \cdots + 1 - d + 1, 0\} - \cdots + (-1)^d \max\{1/2 + \cdots + 1/2 - d + 1, 0\} = 1 - d/2 < 0$, which contradicts the fact that a copula is d-increasing. To see that for $\boldsymbol{u} \in I^d$ there exists a copula C such that $W(\boldsymbol{u}) = C(\boldsymbol{u})$, consider the following cases. First note that if $\boldsymbol{u} = \mathbf{0}$ or $\boldsymbol{u} = \mathbf{1}$, any copula C satisfies $W(\boldsymbol{u}) = C(\boldsymbol{u})$, e.g., take the copula M. Now consider the case where $\boldsymbol{u} \notin \{\mathbf{0}, \mathbf{1}\}$. If $W(\boldsymbol{u}) = 0$ define $A := \{\boldsymbol{v} \in I^d : v_j \in \{0, t_j, 1\}, j \in \{1, \ldots, d\}\}$ where $t_j := \min\{(d-1) u_j / \sum_{k=1}^{d} u_k, 1\}$, $j \in \{1, \ldots, d\}$. Verify that $C'(\boldsymbol{v}) := W(\boldsymbol{v})$ defines a subcopula on A and extend C' to a copula C via Lemma 1.2.3. Then, for each $\boldsymbol{x}$ in the interval $[\mathbf{0}, \boldsymbol{t}]$, which contains $\boldsymbol{u}$, $C(\boldsymbol{x}) = 0 = W(\boldsymbol{u})$. Similarly, for the case where $W(\boldsymbol{u}) > 0$, consider the set A with $t_j := 1 - (1 - u_j)/(d - \sum_{k=1}^{d} u_k)$, $j \in \{1, \ldots, d\}$ and verify that $C'(\boldsymbol{v}) = W(\boldsymbol{v})$ defines a subcopula on A which can be extended to a copula C via Lemma 1.2.3. Then, for each $\boldsymbol{x}$ in the interval $[\boldsymbol{t}, \mathbf{1}]$, which contains $\boldsymbol{u}$, $C(\boldsymbol{x}) = \sum_{j=1}^{d} x_j - d + 1 = W(\boldsymbol{x})$. □

1.2 Definition and basic properties of subcopulas and copulas

From a more theoretical point of view, the set of all copulas is a subset of the set of all continuous functions on I^d and therefore a Banach space with respect to the supremum norm. This Banach space is convex. Also note that the set of all copulas is equicontinuous as well as pointwise bounded, by Proposition 1.2.2 and Theorem 1.2.4, respectively. The Arzelà-Ascoli Theorem, see, e.g., Folland (1999, p. 137), implies that the set of all copulas is compact. Since compact metric spaces are separable, the set of all copulas equipped with the supremum metric contains a countable dense subset. This set can be constructed by means of the following theorem, see Nešlehová (2004, pp. 23) for a proof.

Theorem 1.2.5
Let C be a d-dimensional copula. Then, for any $n \in \mathbb{N}$ and d-dimensional copula C',

$$C_{n,C'}(\boldsymbol{u}) := \sum_{\boldsymbol{i} \in \{1,\ldots,n\}^d} \Delta_{S_{\boldsymbol{i}}} C'(\boldsymbol{\lambda_i}(\boldsymbol{u}))$$

is a d-dimensional copula, where $S_{\boldsymbol{i}} := \prod_{j=1}^{d} [(i_j - 1)/n, i_j/n]$ and

$$\boldsymbol{\lambda_i}(\boldsymbol{u}) := \begin{pmatrix} \max\{\min\{n(u_1 - (i_1 - 1)/n), 1\}, 0\} \\ \vdots \\ \max\{\min\{n(u_d - (i_d - 1)/n), 1\}, 0\} \end{pmatrix}$$

for all $u \in I^d$. Further, for all $\varepsilon > 0$ there exists an $n \in \mathbb{N}$ such that

$$\sup_{\boldsymbol{u} \in I^d} |C(\boldsymbol{u}) - C_{n,C'}(\boldsymbol{u})| < \varepsilon.$$

Since a d-dimensional copula C is a distribution function, it induces a unique probability measure $\mathbb{P}_C$ on I^d equipped with the Borel σ-algebra. This follows by defining the premeasure $\mu_0((\boldsymbol{a}, \boldsymbol{b}]) := \Delta_{(\boldsymbol{a},\boldsymbol{b}]} C$ for all $\boldsymbol{a}, \boldsymbol{b} \in I^d : \boldsymbol{a} \leq \boldsymbol{b}$ on the algebra of all rectangles $(\boldsymbol{a}, \boldsymbol{b}]$, thus $\mathbb{P}_C$ is the unique extension of μ_0 to a measure on I^d, see Folland (1999, p. 31). By the Lebesgue Decomposition, see Folland (1999, p. 91), one has $\mathbb{P}_C((\boldsymbol{a}, \boldsymbol{b}]) = \mu_a((\boldsymbol{a}, \boldsymbol{b}]) + \mu_s((\boldsymbol{a}, \boldsymbol{b}])$, where μ_a is the absolutely continuous, finite, and positive measure with existing Radon-Nikodym derivative with respect to the Lebesgue measure on I^d and $\mu_s := \mathbb{P}_C - \mu_a$ is the finite and positive measure which is singular with respect to the Lebesgue measure on I^d. This motivates the following definition.

Definition 1.2.6
For a copula C and $\boldsymbol{u} \in I^d$, $C(\boldsymbol{u}) = A_C(\boldsymbol{u}) + S_C(\boldsymbol{u})$, where the *absolutely continuous component* with Radon-Nikodym derivative c with respect to the Lebesgue measure on I^d and the *singular*

component of C are defined by

$$A_C(\boldsymbol{u}) := \int_{(\boldsymbol{0},\boldsymbol{u}]} c(\boldsymbol{x})\,d\boldsymbol{x},$$

$$S_C(\boldsymbol{u}) := C(\boldsymbol{u}) - A_C(\boldsymbol{u}),$$

respectively. If $C = A_C$ on I^d, then C is *absolutely continuous*, whereas if $C = S_C$ on I^d, C is *singular*.

By Theorem 1.1.3, $\mathbb{P}_C(\{\boldsymbol{u}\}) = 0$ for all $u \in I^d$. Thus, the above decomposition does not have a pure point component, i.e., copulas have no "atoms". Further, note that a copula is singular if and only if its support, i.e., the complement of the union of all open subsets of I^d with $\mathbb{P}_C$-measure zero, has Lebesgue measure zero. Note that by definition and Identity (1.1), it is readily inferred that $\Pi(\boldsymbol{u}) := \prod_{j=1}^{d} u_j$, $\boldsymbol{u} \in I^d$, is a proper copula. Moreover, it is absolutely continuous. By using $C' = \Pi$ and $C' = M$, respectively, in Theorem 1.2.5, one obtains that both the set of all absolutely continuous and the set of all singular copulas are dense in the set of all copulas.

1.3 Sklar's Theorem

In this section we present the main theorem in copula theory, referred to as *Sklar's Theorem*. This result was a milestone for a new field of mathematics. It originates from a letter to Sklar, in which Fréchet raised the question of determining the relationship between a multidimensional distribution function and its lower dimensional margins. By introducing copulas, Sklar provided an answer to this question for one-dimensional margins. His result was then published by Fréchet as Sklar (1959).

Theorem 1.3.1 (Sklar (1959))

Let H be a distribution function with margins F_j, $j \in \{1, \ldots, d\}$. Then, there exists a copula C such that

$$H(x_1, \ldots, x_d) = C(F_1(x_1), \ldots, F_d(x_d)), \quad \boldsymbol{x} \in \mathbb{R}^d. \tag{1.3}$$

C is uniquely determined on $\prod_{j=1}^{d} \operatorname{ran} F_j$, $j \in \{1, \ldots, d\}$, and given by

$$C(\boldsymbol{u}) = H(F_1^-(u_1), \ldots, F_d^-(u_d)), \quad \boldsymbol{u} \in \prod_{j=1}^{d} \operatorname{ran} F_j.$$

Conversely, given a copula C and univariate distribution functions F_j, $j \in \{1, \ldots, d\}$, H defined by (1.3) is a distribution function with margins F_j, $j \in \{1, \ldots, d\}$.

1.3 Sklar's Theorem

Proof

Consider the first part of the statement and let $\boldsymbol{a}, \boldsymbol{b} \in \mathbb{R}^d$. By Theorem 1.1.3, $F_j(a_j) = F_j(b_j)$ for all $j \in \{1, \ldots, d\}$ implies that $H(\boldsymbol{a}) = H(\boldsymbol{b})$. Therefore, the set $\{((F_1(x_1), \ldots, F_d(x_d)), H(\boldsymbol{x}))^T : \boldsymbol{x} \in \mathbb{R}^d\}$ defines a unique function C' with domain $\prod_{j=1}^d \operatorname{ran} F_j$. By Proposition 1.1.6 (4), C' is given by

$$C'(\boldsymbol{u}) = C'\big(F_1(F_1^-(u_1)), \ldots, F_d(F_d^-(u_d))\big) = H(F_1^-(u_1), \ldots, F_d^-(u_d)), \quad \boldsymbol{u} \in \prod_{j=1}^d \operatorname{ran} F_j \quad (1.4)$$

By the properties of H, C' is a subcopula with domain $\prod_{j=1}^d \operatorname{ran} F_j$. If the margins F_j, $j \in \{1, \ldots, d\}$, are continuous, then C' is the unique copula C as claimed. Otherwise, extend C' via Lemma 1.2.3 to obtain a copula C as claimed. For the converse statement, let $\boldsymbol{U} \sim C$ and define $\boldsymbol{X} := (F_1^-(U_1), \ldots, F_d^-(U_d))^T$. Proposition 1.1.6 (5) implies

$$\mathbb{P}(\boldsymbol{X} \le \boldsymbol{x}) = \mathbb{P}(F_1^-(U_1) \le x_1, \ldots, F_d^-(U_d) \le x_d) = \mathbb{P}(U_1 \le F_1(x_1), \ldots, U_d \le F_d(x_d))$$
$$= C(F_1(x_1), \ldots, F_d(x_d)). \qquad \square$$

Remark 1.3.2

(1) On the one hand, Sklar's Theorem tells us that we can decompose any given multivariate distribution function into its margins and a copula. By this decomposition, copulas allow us to study multivariate distributions functions independently of the margins. This is of particular interest in statistics. On the other hand, Sklar's Theorem provides a tool for constructing multivariate distributions and is therefore often used for sampling purposes.

(2) To see that Decomposition (1.3) is not necessarily unique if the margins F_j, $j \in \{1, \ldots, d\}$, are not continuous, consider the random vector $(X_1, X_2)^T$ following the bivariate Bernoulli distribution, given by $\mathbb{P}(X_1 = i, X_2 = j) = 1/4$, $i, j \in \{0, 1\}$. Since $\mathbb{P}(X_1 = 0) = \mathbb{P}(X_2 = 0) = 1/2$, $\operatorname{ran} F_1 = \operatorname{ran} F_2 = \{0, 1/2, 1\}$. Now observe that any copula C with $C(1/2, 1/2) = 1/4$ satisfies $H(x_1, x_2) = C(F_1(x_1), F_2(x_2))$ for all $x_i \in \bar{\mathbb{R}}$, $i \in \{1, 2\}$. Thus, on the interior of I^2, a copula associated to F_1, F_2, and H is only uniquely determined at the single point $(1/2, 1/2)^T$, where it should equal $1/4$. Two different copulas that share this property are, e.g., the copulas $C_1(u_1, u_2) = \Pi(u_1, u_2)$ and $C_2(u_1, u_2) = \min\{u_1, u_2, (\delta(u_1) + \delta(u_2))/2\}$, the latter being a *diagonal copula* with diagonal $\delta(t) = t^2$, $t \in I$, see Nelsen (2007, pp. 84).

27

1 Copulas

1.4 Random vectors and copulas

In this section we explore the connection between random vectors and copulas. By Sklar's Theorem, a random vector $\boldsymbol{X} = (X_1, \ldots, X_d)^T \sim H$ with continuous univariate margins F_j, $j \in \{1, \ldots d\}$, possesses an associated copula, the *copula of* $\boldsymbol{X}$, equivalently, the *copula of* H. This copula is simply the distribution function of the random vector $(F_1(X_1), \ldots, F_d(X_d))^T$ as the following lemma shows.

Lemma 1.4.1
Let $X_j \sim F_j$, with F_j continuous, $j \in \{1, \ldots, d\}$. Then, the copula of $\boldsymbol{X}$ is the distribution function of $(F_1(X_1), \ldots, F_d(X_d))^T$.

Proof
By Proposition 1.1.7 (1), $F_j(X_j) \sim \mathrm{U}[0,1]$. Thus, $F_j(X_j)$ is continuously distributed for all $j \in \{1, \ldots, d\}$. This implies that the distribution function of $(F_1(X_1), \ldots, F_d(X_d))^T$ is $\mathbb{P}(F_1(X_1) < u_1, \ldots, F_d(X_d) < u_d)$. By Proposition 1.1.6 (5), this equals $\mathbb{P}(X_1 < F_1^-(u_1), \ldots, X_d < F_d^-(u_d))$, thus $\mathbb{P}(X_1 \leq F_1^-(u_1), \ldots, X_d \leq F_d^-(u_d))$. By Sklar's Theorem, this equals $C(\boldsymbol{u})$ for all $\boldsymbol{u} \in \prod_{j=1}^d \mathrm{ran}\, F_j$. Continuity of F_j, $j \in \{1, \ldots, d\}$, establishes the claim. □

The following theorem summarizes properties of copulas and their corresponding random variables.

Theorem 1.4.2
Let H be a d-dimensional distribution function with continuous margins F_j, $j \in \{1, \ldots, d\}$, and copula C. Further, let $\boldsymbol{X} \sim H$.

(1) If T_j is strictly increasing on $\mathrm{ran}\, X_j$, $j \in \{1, \ldots, d\}$, then $(T_1(X_1), \ldots, T_d(X_d))^T$ also has copula C, i.e., copulas are invariant under strictly increasing transformations on the ranges of the underlying random variables.

(2) $H(\boldsymbol{x}) = \prod_{j=1}^d F_j(x_i)$, $\boldsymbol{x} \in \mathbb{R}^d$, if and only if $C(\boldsymbol{u}) = \Pi(\boldsymbol{u})$, $\boldsymbol{u} \in I^d$, i.e., the random variables X_j, $j \in \{1, \ldots, d\}$, are independent if and only if the underlying copula is Π, the so-called *independence copula*.

(3) If $d = 2$, then $X_2 = T(X_1)$ almost surely with $T := F_2^- \circ (1 - F_1)$ if and only if $C(\boldsymbol{u}) = W(\boldsymbol{u})$, $\boldsymbol{u} \in I^2$; i.e., X_2 is almost surely a strictly decreasing function in X_1 if and only if the underlying copula is the lower Fréchet bound.

(4) $X_j = T_j(X_1)$ almost surely with $T_j := F_j^- \circ F_1$, $j \in \{2, \ldots, d\}$, if and only if $C(\boldsymbol{u}) = M(\boldsymbol{u})$, $\boldsymbol{u} \in I^d$; i.e., the random variables X_j, $j \in \{2, \ldots, d\}$, are almost surely strictly increasing

functions in X_1 if and only if the underlying copula is the upper Fréchet bound.

Proof

(1) Since T_j is strictly increasing on $\operatorname{ran} X_j$, it has at most countably many discontinuities there. Without loss of generality, assume T_j to be right-continuous at its discontinuities and apply Proposition 1.1.6 (5) to see that $F_{T_j(X_j)}(x) := \mathbb{P}(T_j(X_j) \leq x) = \mathbb{P}(X_j \leq T_j^-(x)) = F_j(T_j^-(x))$. Since X_j is continuously distributed and T_j^- is continuous on $\operatorname{ran} T_j(X_j)$, by Proposition 1.1.6 (2), $F_{T_j(X_j)}$ is continuous. By Proposition 1.1.6 (3), $\mathbb{P}\bigl(F_{T_1(X_1)}(T_1(X_1)) \leq u_1, \ldots, F_{T_d(X_d)}(T_d(X_d)) \leq u_d\bigr) = \mathbb{P}\bigl(F_1\bigl(T_1^-(T_1(X_1))\bigr) \leq u_1, \ldots, F_d\bigl(T_d^-(T_d(X_d))\bigr) \leq u_d\bigr) = \mathbb{P}(F_1(X_1) \leq u_1, \ldots, F_d(X_d) \leq u_d)$, which, by Lemma 1.4.1, equals $C(\boldsymbol{u})$. Therefore, the claim follows by continuity of $F_j(T_j^-(x))$ and a further application of Lemma 1.4.1.

(2) Equation (1.3) implies $C(F_1(x_1), \ldots, F_d(x_d)) = H(x_1, \ldots, x_d) = \prod_{j=1}^d F_j(x_j)$, $\boldsymbol{x} \in \mathbb{R}^d$, if and only if $C(\boldsymbol{u}) = \prod_{j=1}^d u_j$, $\boldsymbol{u} \in I^d$.

(3) For the "only if" part, note that $F_2^-(1-F_1(X_1)) \sim F_2$, by Proposition 1.1.7 (1), (2). Equation (1.4) therefore implies $C_{X_1, X_2}(u_1, u_2) = C_{X_1, F_2^-(1-F_1(X_1))}(u_1, u_2) = \mathbb{P}(X_1 \leq F_1^-(u_1), F_2^-(1-F_1(X_1)) \leq F_2^-(u_2))$. By Proposition 1.1.7 (2), this equals $\mathbb{P}\bigl(X_1 \leq F_1^-(u_1), 1 - F_1(X_1) \leq F_2(F_2^-(u_2))\bigr)$, which, by the second part of Proposition 1.1.6 (4) equals $\mathbb{P}(X_1 \leq F_1^-(u_1), 1 - F_1(X_1) \leq u_2) = \mathbb{P}(F_1^-(1 - u_2) \leq X_1 \leq F_1^-(u_1)) = \max\{u_1 + u_2 - 1, 0\}$. Now consider the "if" part. By definition, W puts no mass on $\{\boldsymbol{u} \in I^2 : u_1 + u_2 < 1\}$ and since, for $\boldsymbol{U} \sim W$, $\mathbb{P}(U_1 > u_1, U_2 > u_2) = 1 - u_1 - u_2 + W(u_1, u_2) = 0$ on $\{\boldsymbol{u} \in I^2 : u_1 + u_2 > 1\}$, W also puts no mass on $\{\boldsymbol{u} \in I^2 : u_1 + u_2 > 1\}$. Since F_2 is almost surely strictly increasing on $\operatorname{ran} X_2$, it follows that $\mathbb{P}\bigl(X_2 = F_2^-(1 - F_1(X_1))\bigr) = \mathbb{P}\bigl(F_2(X_2) = F_2\bigl(F_2^-(1 - F_1(X_1))\bigr)\bigr)$, which, by the second part of Proposition 1.1.6 (4), equals $\mathbb{P}(F_1(X_1) + F_2(X_2) - 1 = 0) = 1$.

(4) Similarly as (3), where we note that M puts no mass on the complement of $\{\boldsymbol{u} \in I^d : u_1 = \cdots = u_d\}$.

$\square$

Remark 1.4.3

(1) Lemma 1.4.1 and Theorem 1.4.2 go back to Schweizer and Wolff (1981). These authors establish a link between copula theory and the investigation of dependencies between random variables for which copulas are mostly applied today. Theorem 1.4.2 (1) implies that copulas precisely capture the dependence structure between random variables, independently of the margins. According to Fisher (1997), this is one of the reasons why copulas are so important.

1 Copulas

(2) The random variables in Theorem 1.4.2 (3) correspond to *perfect negative dependence* and are called *countermonotone*; the ones in Part (4) correspond to *perfect positive dependence* and are called *comonotone*.

Note that the following result replaces Theorem 1.4.2 (1) when the margins are not continuous.

Theorem 1.4.4
Let $X_j \sim F_j$, $j \in \{1, \ldots, d\}$, with copula C. If T_j is continuous and strictly increasing on ran X_j, $j \in \{1, \ldots, d\}$, then $(T_1(X_1), \ldots, T_d(X_d))^T$ has copula C.

Proof
First consider the margins of $(T_1(X_1), \ldots, T_d(X_d))^T$. Let $j \in \{1, \ldots, d\}$. Since T_j is strictly increasing on ran X_j, $F_j(T_j^-(x)) = \mathbb{P}(X_j \leq T_j^-(x)) = \mathbb{P}(T(X_j) \leq T_j(T_j^-(x)))$. By Proposition 1.1.6 (4), this equals $\mathbb{P}(T_j(X_j) \leq x) =: F_{T_j(X_j)}(x)$ for all $x \in \operatorname{ran} T_j$. Since $T_j(X_j)$ puts no mass outside ran T_j, one has $F_{T_j(X_j)}(x) = F_j(T_j^-(x))$, $x \in \bar{\mathbb{R}}$. On ran X_j, $F_{T_j(X_j)}^- = (F_j \circ T_j^-)^-$, which equals $(T_j^-)^- \circ F_j^-$, see Proposition 1.1.6 (2), (6). Since T_j is continuous and strictly increasing on ran X_j, $T_j^- = T_j^{-1}$, thus $F_{T_j(X_j)}^- = T_j \circ F_j^-$ on ran X_j. This also holds on $\mathbb{R}$ because X_j puts not mass outside ran X_j. With this result at hand, we now show that C combines $F_{T_j(X_j)}$ to the distribution function of $(T_1(X_1), \ldots, T_d(X_d))^T$ to conclude the claim by Sklar's Theorem. For this, let $(U_1, \ldots, U_d)^T \sim C$ and note that Proposition 1.1.6 (5) implies $C(F_{T_1(X_1)}(x_1), \ldots, F_{T_d(X_d)}(x_d)) = \mathbb{P}(U_1 \leq F_{T_1(X_1)}(x_1), \ldots, U_d \leq F_{T_d(X_d)}(x_d)) = \mathbb{P}(F_{T_1(X_1)}^-(U_1) \leq x_1, \ldots, F_{T_d(X_d)}^-(U_d) \leq x_d)$. By the above result, this equals $\mathbb{P}(T_1(F_1^-(U_1)) \leq x_1, \ldots, T_d(F_d^-(U_d)) \leq x_d)$, which, by Proposition 1.1.7 (2), equals $\mathbb{P}(T_1(X_1) \leq x_1, \ldots, T_d(X_d) \leq x_d)$. □

1.5 Survival copulas

The construction of copulas is one of the most active fields of copula research. By Sklar's Theorem, having a repertoire of copulas at hand allows one to construct multivariate distributions which in turn creates flexibility for modeling purposes. Besides investigation of dependencies, constructing multivariate distribution functions is another reason for the popularity of copulas, as mentioned by Fisher (1997). Since a large proportion of the existing copula theory deals with the bivariate case, there is a demand for principles to construct multivariate copula families. For a given random vector, one simple construction principle to obtain another copula is that of survival copulas, based on the following analog of Sklar's Theorem in terms of survival functions.

1.6 Symmetries of copulas

Theorem 1.5.1 (Sklar (1959))
Let $\bar{H}$ be a survival function with margins $\bar{F}_j$, $j \in \{1,\ldots,d\}$. Then, there exists a copula $\hat{C}$ such that

$$\bar{H}(x_1,\ldots,x_d) = \hat{C}(\bar{F}_1(x_1),\ldots,\bar{F}_d(x_d)), \quad \boldsymbol{x} \in \mathbb{R}^d. \tag{1.5}$$

$\hat{C}$ is uniquely determined on $\prod_{j=1}^d \operatorname{ran} \bar{F}_j$, $j \in \{1,\ldots,d\}$, and given by

$$\hat{C}(\boldsymbol{u}) = \bar{H}(\bar{F}_1^-(u_1),\ldots,\bar{F}_d^-(u_d)), \quad \boldsymbol{u} \in \prod_{j=1}^d \operatorname{ran} \bar{F}_j, \tag{1.6}$$

with $\bar{F}_j^-(y) := F_j^-(1-y)$, $j \in \{1,\ldots,d\}$. Conversely, given a copula $\hat{C}$ and univariate survival functions $\bar{F}_j$, $j \in \{1,\ldots,d\}$, $\bar{H}$ defined by (1.5) is a survival function with margins $\bar{F}_j$, $j \in \{1,\ldots,d\}$.

In terms of Sklar's Theorem based on survival functions, the survival copula links a multivariate survival function to its one-dimensional survival functions. A copula $\hat{C}$ defined by (1.5) is therefore called *survival copula* of $\bar{H}$.

Remark 1.5.2
Note that $\boldsymbol{U} \sim C$ if and only if $(1-U_1,\ldots,1-U_d)^T \sim \hat{C}$. By the Poincaré-Sylvester sieve formula, there is a direct connection between $\hat{C}$ and C, given by

$$\hat{C}(\boldsymbol{u}) = 1 - \sum_{j=1}^d (-1)^{j-1} \sum_{1 \leq i_1 < \cdots < i_j \leq d} C^{(i_1 \ldots i_j)}(1-u_{i_1},\ldots,1-u_{i_j}),$$

where $C^{(i_1 \ldots i_j)}$ denotes the $(i_1,\ldots,i_j)$-th margin of C, obtained by letting $u_k = 1$ for all $k \notin \{i_1,\ldots,i_j\}$.

1.6 Symmetries of copulas

In this section we briefly present two concepts of symmetries of random variables. For other kinds of symmetries, see, e.g., Nelsen (2007, p. 36).

Definition 1.6.1
A d-dimensional random vector $\boldsymbol{X}$ is called

(1) *radially symmetric* (for $d = 1$ simply *symmetric*) about $\boldsymbol{a} \in \mathbb{R}^d$ if $\boldsymbol{X} - \boldsymbol{a}$ and $\boldsymbol{a} - \boldsymbol{X}$ are equally distributed;

31

1 Copulas

(2) *exchangeable* if $\boldsymbol{X} = (X_1, \ldots, X_d)^T$ and $(X_{j_1}, \ldots, X_{j_d})^T$ are equally distributed for all permutations $(j_1, \ldots, j_d)$ of $\{1, \ldots, d\}$.

For a random vector $\boldsymbol{X}$, having independent and identically distributed components implies exchangeability. Exchangeability in turn implies having identically distributed components. To see that the corresponding converse statements fail in general, let $\boldsymbol{X}$ have equal margins. For the converse of the first statement, combine the components of $\boldsymbol{X}$ with any symmetric copula which is not Π, e.g., a Clayton copula, see Section 2.4. For the converse of the second statement, combine the components of $\boldsymbol{X}$ with any non-symmetric copula, e.g., a Marshall-Olkin copula with $\alpha_1 \neq \alpha_2$, see Section A.2.

The following proposition is a straightforward multivariate generalization of the results in Nelsen (2007, pp. 37).

Proposition 1.6.2
Let H be a distribution function with continuous margins F_j, $j \in \{1, \ldots, d\}$, and copula C. Further, let $\boldsymbol{X} \sim H$.

(1) If X_j is symmetric about a_j, $j \in \{1, \ldots, d\}$, then $\boldsymbol{X}$ is radially symmetric about $\boldsymbol{a}$ if and only if $C = \hat{C}$.

(2) $\boldsymbol{X}$ is exchangeable if and only if $F_1 = \cdots = F_d$ and $C(u_1, \ldots, u_d) = C(u_{j_1}, \ldots, u_{j_d})$ for all permutations $(j_1, \ldots, j_d)$ of $\{1, \ldots, d\}$ and $\boldsymbol{u} \in I^d$.

Another question is to ask if radial symmetry and exchangeability are comparable. This is not the case. To see that exchangeability does not imply being radially symmetric, consider a Clayton copula, see, e.g., Section 2.4. For the converse statement, a radially symmetric copula which is not exchangeable can be constructed via $C(u_1, u_2) := u_1 u_2 + f(u_1)g(u_2)$ where f and g are continuously differentiable on I with $f'(u)g'(v) \geq -1$ for all $u, v \in I$ and $f(0) = g(0) = f(1) = g(1) = 0$. Now if f and g are symmetric about $1/2$, then C is radially symmetric, however, if $f \neq g$, then C is not symmetric in general, e.g., take $f(t) = t(1-t)$ and $g(t) = \sin(\pi t)/\pi$.

1.7 Measures of association

It is often desirable to measure the degree of dependence between random variables by a real number. Such *measures of association* were originally studied in the bivariate case and they are still more popular than their possible multivariate generalizations. For this reason, we only treat bivariate measures of association in the sequel, with focus on the linear correlation coefficient, measures of dependence, measures of concordance, and the coefficients of tail dependence.

1.7.1 Correlation

Pearson's linear correlation coefficient is the most widely used measure of association. It is defined as follows.

Definition 1.7.1

The *correlation*, or *correlation coefficient*, between two random variables X_1 and X_2 with $\mathbb{E}[X_j^2] < \infty$, $j \in \{1, 2\}$, is defined by

$$\rho := \rho_{12} := \rho_{X_1, X_2} := \frac{\mathrm{Cov}[X_1, X_2]}{\sqrt{\mathrm{Var}[X_1]}\sqrt{\mathrm{Var}[X_2]}} = \frac{\mathbb{E}[(X_1 - \mathbb{E}[X_1])(X_2 - \mathbb{E}[X_2])]}{\sqrt{\mathbb{E}[(X_1 - \mathbb{E}[X_1])^2]}\sqrt{\mathbb{E}[(X_2 - \mathbb{E}[X_2])^2]}}.$$

Some well-known properties of the correlation coefficient are summarized in the following proposition.

Proposition 1.7.2

Let X_1 and X_2 be two random variables with $\mathbb{E}[X_j^2] < \infty$, $j \in \{1, 2\}$. Then,

(1) $-1 \leq \rho \leq 1$;

(2) $|\rho| = 1$ if and only if there exist real numbers $a \neq 0$ and b such that $X_2 = aX_1 + b$ almost surely, i.e., X_1 and X_2 are *perfectly linearly dependent*. If $\rho = -1$, then $a < 0$ and if $\rho = 1$ then $a > 0$.

(3) If X_1 and X_2 are independent, then $\rho = 0$. However, the converse statement is false in general.

(4) Correlation is invariant under strictly increasing linear transformations on the ranges of the underlying random variables. However, it is in general not invariant under nonlinear such transformations.

According to Embrechts et al. (2002), the popularity of correlation is due to the fact that the correlation coefficient is often straightforward to calculate and easy to manipulate under linear operations. Further, it is a natural measure in the context of elliptical distributions, see Section 1.9. In the non-elliptical world, however, it can be quite misleading to use the correlation coefficient as a measure of association, see Embrechts et al. (2002) or McNeil et al. (2005, pp. 201) for further details. The main problem is that it also depends on the margins of the random variables under consideration. This already affects the range of possible values for the

1 Copulas

correlation coefficient. One usually cannot expect to attain the whole range [-1,1], see McNeil et al. (2005, p. 205) for an example involving two log-normally distributed random variables. Another consequence is that correlation is not invariant under strictly increasing transformations in general. By Proposition 1.4.2 (1) such invariance is a copula property. However, since correlation also depends on the margins, it is not possible to study the underlying dependence independently of the margins. This is also undesirable from a statistical point of view, since transforming the given data with the margins may change the correlation coefficient. Moreover, the margins even have an influence on the existence of the correlation coefficient since it is not defined if the second moments of the underlying random variables do not exist.

Concerning the deficiencies of correlation, a desirable property of a measure of association for two continuously distributed random variables $X_j \sim F_j$, $j \in \{1,2\}$ is that it should only depend on their copula. According to Theorem 1.4.2 (1), it suffices to study such measures of association in terms of *ranks*, i.e., the random variables $F_j(X_j)$, $j \in \{1,2\}$. These measures are therefore also called *rank correlation measures*. Since they are functionals of the copula, rank correlation measures have many desirable properties and can be used to fit copulas to empirical data. In the sequel, we present two classes of such measures, measures of dependence and measures of concordance. Due to Sklar's Theorem, the restriction to continuously distributed random variables is natural in order to have a unique copula for studying dependence. For generalizations to not necessarily continuously distributed random variables, we refer the interested reader to Nešlehová (2004) and Nešlehová (2007).

1.7.2 Measures of dependence

The term "measure of dependence" already appears in Hoeffding (1940). A set of desirable properties of such measures is introduced in Rényi (1959) and partly modified, as well as combined with the concept of copulas, by Schweizer and Wolff (1981). The resulting axioms are again slightly modified in Embrechts et al. (2002). In simple terms, measures of dependence capture the degree of dependence between independence and perfect dependence, where the latter comprises perfect negative and perfect positive dependence, given by the copulas W and M, respectively. The following definition, as well as examples, can be found, e.g., in Nelsen (2007, pp. 207).

Definition 1.7.3

A measure of association $\delta = \delta_{12} = \delta_{X_1,X_2} = \delta_C$ between two continuously distributed random variables X_1 and X_2 with copula C is a *measure of dependence* if the following properties hold:

(1) δ is defined for every pair X_1, X_2 of continuously distributed random variables;

(2) $0 \leq \delta \leq 1$;

(3) $\delta_{12} = \delta_{21}$;

(4) $\delta_{12} = \delta_\Pi = 0$ if and only if X_1 and X_2 are independent;

(5) $\delta_{12} = 1$ if and only if X_1 and X_2 are comonotone or countermonotone;

(6) if T_j is a strictly increasing function on ran X_j, $j \in \{1,2\}$, then $\delta_{T_1(X_1),T_2(X_2)} = \delta_{X_1,X_2}$;

(7) if $(C_n)_{n \in \mathbb{N}}$ is a sequence of copulas which converges pointwise to C, then $\lim_{n \to \infty} \delta_{C_n} = \delta_C$.

Note that measures of dependence can be constructed by finding suitable metrics on the set of copulas, see Nešlehová (2004, p. 57).

As Embrechts et al. (2002) point out, the disadvantage of these measures is that they cannot differentiate between positive and negative dependence. To be more precise, a measure which indicates negative dependence by negative values requires the property that $\delta_{-X_1,X_2} = -\delta_{X_1,X_2}$. This contradicts Definition 1.7.3 (4) which can be seen by taking $(X_1, X_2)^T$ to be uniformly distributed on the unit circle. Then, $\delta_{-X_1,X_2} = \delta_{X_1,X_2}$, but also $\delta_{-X_1,X_2} = -\delta_{X_1,X_2}$. Thus, $\delta_{X_1,X_2} = 0$ although X_1 and X_2 are not independent. Further, it is often not clear how to estimate measures of dependence. For this reason, the concept of measures of concordance is introduced.

1.7.3 Measures of concordance

The Fréchet bounds suggest a pointwise partial order $\preceq$ on the set of copulas, defined as follows, see Joe (1997, p. 21) and Joe (1997, pp. 36). The presented notion is used interchangeably for the underlying random variables.

Definition 1.7.4

Let C_i be a d-dimensional copula, $i \in \{1,2\}$. Then C_1 is *less positive lower orthant dependent than* C_2, equivalently C_2 is *more positive lower orthant dependent than* C_1, denoted by $C_1 \preceq C_2$, if $C_1(\boldsymbol{u}) \leq C_2(\boldsymbol{u})$ for all $\boldsymbol{u} \in I^d$. If $C_1 = \Pi$, C_2 is called *positive lower orthant dependent*.

The following definition of a measure of concordance, see Scarsini (1984) and Nelsen (2007, pp. 168), reflects desirable properties of measures of association. In contrast to Definition 1.7.3, measures of concordance can differentiate between positive and negative dependence. Due to Definition 1.7.5 (6), the partial order $\preceq$ of the set of bivariate copulas is called *concordance*

1 Copulas

ordering and C_1 is said to be *less concordant than* C_2, equivalently C_2 is said to be *more concordant than* C_1; similarly for the underlying pairs of random variables.

Definition 1.7.5
A measure of association $\kappa := \kappa_{12} := \kappa_{X_1,X_2} := \kappa_C$ between two continuously distributed random variables X_1 and X_2 with copula C is a *measure of concordance* if the following properties hold:

(1) κ is defined for every pair X_1, X_2 of continuously distributed random variables;

(2) $-1 \leq \kappa \leq 1$, $\kappa_W = -1$, and $\kappa_M = 1$;

(3) $\kappa_{12} = \kappa_{21}$;

(4) if X_1 and X_2 are independent, then $\kappa_{12} = \kappa_\Pi = 0$;

(5) $\kappa_{-X_1,X_2} = -\kappa_{X_1,X_2}$;

(6) if C_1 and C_2 are bivariate copulas such that $C_1 \preceq C_2$, then $\kappa_{C_1} \leq \kappa_{C_2}$;

(7) if $(C_n)_{n \in \mathbb{N}}$ is a sequence of bivariate copulas which converges pointwise to C, then $\lim_{n \to \infty} \kappa_{C_n} = \kappa_C$.

As a consequence of Definition 1.7.5 and Theorem 1.4.2, measures of concordance share the following properties, see Scarsini (1984).

Proposition 1.7.6
Let κ be a measure of concordance for two continuously distributed random variables X_1 and X_2.

(1) If X_2 is almost surely a strictly decreasing function in X_1, then $\kappa_{12} = \kappa_W = -1$;

(2) if X_2 is almost surely a strictly increasing function in X_1, then $\kappa_{12} = \kappa_M = 1$;

(3) if T_j is a strictly increasing function on ran X_j, $j \in \{1,2\}$, then $\kappa_{T_1(X_1),T_2(X_2)} = \kappa_{X_1,X_2}$.

Note that measures of concordance can also be constructed by finding suitable metrics on the set of copulas, see Nešlehová (2004, p. 62).

We now consider two specific measures of concordance, Spearman's rho and Kendall's tau.

Definition 1.7.7
Let $X_j \sim F_j$, $j \in \{1,2\}$, be continuously distributed random variables with copula C. Then *Spearman's rho* is defined by

$$\rho_S := \rho_{S,12} := \rho_{S,X_1,X_2} := \rho_{S,C} := \rho_{F_1(X_1),F_2(X_2)}.$$

1.7 Measures of association

If $(X_1', X_2')^T$ is an independent and identically distributed copy of $(X_1, X_2)^T$, then *Kendall's tau* is defined by

$$\tau := \tau_{12} := \tau_{X_1, X_2} := \tau_C := \mathbb{E}[\operatorname{sgn}((X_1 - X_1')(X_2 - X_2'))],$$

where $\operatorname{sgn}(x) := \mathbb{1}_{(0, \infty)}(x) - \mathbb{1}_{(-\infty, 0)}(x)$.

For a *random sample* $(x_{1i}, x_{2i})^T$, $i \in \{1, \ldots, n\}$, of $(X_1, X_2)^T$, i.e., realizations of independent copies of $(X_1, X_2)^T$, note that Pearson's correlation coefficient as well as Spearman's and Kendall's rank correlation coefficients have obvious estimates, also called *sample versions* and denoted by $\hat\rho$, $\hat\rho_S$, and $\hat\tau$, respectively. For example, the sample version of Kendall's tau is given by

$$\hat\tau = \frac{1}{\binom{n}{2}} \sum_{1 \leq i_1 < i_2 \leq n} \operatorname{sgn}((x_{1i_1} - x_{1i_2})(x_{2i_1} - x_{2i_2})).$$

Note that realizations $(x_1, x_2)^T$ and $(x_1', x_2')^T$ of two independent copies $(X_1, X_2)^T$ and $(X_1', X_2')^T$ of continuously distributed random variables are *concordant* if $(x_1 - x_1')(x_2 - x_2') > 0$ and *disconcordant* if $(x_1 - x_1')(x_2 - x_2') < 0$. The following theorem shows that Spearman's rho and Kendall's tau can be expressed in terms of the probability of concordance minus the probability of discordance of some underlying random variables. Further, it explains the notation $\rho_{S,C}$ and τ_C in Definition 1.7.7, i.e., that ρ_S and τ only depend on the copula of the underlying random variables.

Theorem 1.7.8
Let $(X_1, X_2)^T$ and $(X_1', X_2')^T$ be independent vectors of continuously distributed random variables with identical margins F_1, for X_1 and X_1', and F_2, for X_2 and X_2', and copulas C and C', respectively. Let $Q_{CC'} := \mathbb{P}((X_1 - X_1')(X_2 - X_2') > 0) - \mathbb{P}((X_1 - X_1')(X_2 - X_2') < 0)$. Then

$$Q_{CC'} = 4 \int_{I^2} C(\boldsymbol{u})\, dC'(\boldsymbol{u}) - 1.$$

$Q_{CC'}$ is symmetric and increasing in C and C', and $Q_{\hat{C}\hat{C}'} = Q_{CC'}$. Further,

$$\rho_S = 3 Q_{C\Pi} = 12 \int_{I^2} C(\boldsymbol{u})\, d\boldsymbol{u} - 3,$$

$$\tau = Q_{CC} = 4 \int_{I^2} C(\boldsymbol{u})\, dC(\boldsymbol{u}) - 1. \quad (1.7)$$

Proof
By continuity of the random variables considered, $\mathbb{P}((X_1 - X_1')(X_2 - X_2') < 0) = 1 - \mathbb{P}((X_1 - X_1')(X_2 - X_2') > 0)$. Thus, $Q_{CC'} = 2\mathbb{P}((X_1 - X_1')(X_2 - X_2') > 0) - 1$. Now $\mathbb{P}((X_1 - X_1')(X_2 - X_2') >$

1 Copulas

$0) = \mathbb{P}(X_1 < X_1', X_2 < X_2') + \mathbb{P}(X_1 > X_1', X_2 > X_2')$. Both summands can be evaluated by conditioning on either $(X_1, X_2)^T$ or $(X_1', X_2')^T$, leading to

$$\mathbb{P}(X_1 < X_1', X_2 < X_2') = \mathbb{P}(X_1 > X_1', X_2 > X_2') = \int_{I^2} C'(\boldsymbol{u})\, dC(\boldsymbol{u}) = \int_{I^2} C(\boldsymbol{u})\, dC'(\boldsymbol{u})$$

and the first claim, as well as symmetry of $Q_{CC'}$ in C and C' follow. By symmetry in turn, it is easy to see that $Q_{CC'}$ is increasing in C and C'. By Remark 1.5.2, it is straightforward to verify that $Q_{\hat{C}\hat{C}'} = 4\mathbb{E}[\hat{C}(1-U_1', 1-U_2')] - 1 = 4\mathbb{E}[C(U_1', U_2')] - 1 = Q_{CC'}$, where $(U_1', U_2')^T \sim C'$. Now let X_1 and X_2 be two continuously distributed random variables with copula C and let $(U_1, U_2)^T \sim C$. Then,

$$\rho_S = \frac{\mathbb{E}[U_1 U_2] - \mathbb{E}[U_1]\mathbb{E}[U_2]}{\sqrt{\text{Var}[U_1]}\sqrt{\text{Var}[U_2]}} = 12\mathbb{E}[U_1 U_2] - 3 = 12\int_{I^2} C(\boldsymbol{u})\, d\boldsymbol{u} - 3 = 3Q_{C\Pi},$$

$$\tau = \mathbb{P}((X_1 - X_1')(X_2 - X_2') > 0) - \mathbb{P}((X_1 - X_1')(X_2 - X_2') < 0) = Q_{CC'}.$$

□

By Definition 1.7.7 and the properties of the function Q, it can be verified that Spearman's rho and Kendall's tau satisfy all properties of a measure of concordance as stated in Definition 1.7.5. Additionally, the following properties hold.

Proposition 1.7.9

Let κ stand for either Spearman's rho or Kendall's tau of two continuously distributed random variables $X_j \sim F_j$, $j \in \{1,2\}$, with common distribution function H and copula C. Then, $\kappa = -1$ implies $C = W$ and $\kappa = 1$ implies $C = M$.

Proof

By Theorem 1.7.8, the statement for ρ_S is trivial. Hence, consider τ. By definition, $\tau = 1$ means $\mathbb{P}((X_1 - X_1')(X_2 - X_2') > 0) = 1$, where $(X_1', X_2')^T$ is an independent and identically distributed copy of $(X_1, X_2)^T$. This can be identified with two realizations of $(X_1, X_2)^T$ in the product space $(\Omega \times \Omega, \mathcal{F} \otimes \mathcal{F}, \mathbb{P} \times \mathbb{P})$, thus $(\mathbb{P} \times \mathbb{P})(\{(\omega_1, \omega_2) \in \Omega \times \Omega : (X_1(\omega_1) - X_1(\omega_2))(X_2(\omega_1) - X_2(\omega_2)) > 0\}) = 1$. For some $x_j \in \mathbb{R}$, define $A_j := \{\omega \in \Omega : X_j(\omega) \leq x_j\}$, $j \in \{1,2\}$, and assume $\mathbb{P}(A_1) \leq \mathbb{P}(A_2)$, i.e., $F_1(x_1) \leq F_2(x_2)$. If $\mathbb{P}(A_1 \backslash A_2) > 0$, then $\mathbb{P}(A_2 \backslash A_1) > 0$. Thus, $(\mathbb{P} \times \mathbb{P})((A_1 \backslash A_2) \times (A_2 \backslash A_1)) = \mathbb{P}(A_1 \backslash A_2)\mathbb{P}(A_2 \backslash A_1) > 0$. But on $(A_1 \backslash A_2) \times (A_2 \backslash A_1)$, $(X_1(\omega_1) - X_1(\omega_2))(X_2(\omega_1) - X_2(\omega_2)) < 0$, a contradiction. Thus, $\mathbb{P}(A_1 \backslash A_2) = 0$ and therefore $\mathbb{P}(A_1 \cap A_2) = \mathbb{P}(A_1)$. Equivalently, $H(x_1, x_2) = F_1(x_1)$. By the same reasoning, $\mathbb{P}(A_1) \geq \mathbb{P}(A_2)$, i.e., $F_1(x_1) \geq F_2(x_2)$, implies $\mathbb{P}(A_1 \cap A_2) = \mathbb{P}(A_2)$, i.e., $H(x_1, x_2) = F_2(x_2)$. Thus, $\tau = 1$ implies $C = M$. For $\tau = -1$ and $C = W$, proceed similarly.

□

Note that the computation of τ is often simplified by the identity

$$\int_{I^2} C(\boldsymbol{u})\, dC'(\boldsymbol{u}) = \frac{1}{2} - \int_{I^2} D_1\, C(\boldsymbol{u})\, D_2\, C'(\boldsymbol{u})\, d\boldsymbol{u}, \tag{1.8}$$

see Li et al. (2002). Finally, let us note that there are other measures of concordance not discussed here, including Gini's gamma and Blomqvist's beta, see Nelsen (2007, pp. 180).

1.7.4 Tail dependence

Tail dependence measures the extremal dependence between two random variables, i.e., the strength of dependence in the tails of their bivariate distribution.

Definition 1.7.10

Let $X_j \sim F_j$, $j \in \{1, 2\}$, be continuously distributed random variables. The *lower tail-dependence coefficient*, respectively the *upper tail-dependence coefficient*, of X_1 and X_2 are defined as

$$\lambda_l := \lambda_{l,12} := \lambda_{l,X_1,X_2} := \lambda_{l,C} := \lim_{t\downarrow 0} \mathbb{P}(X_2 \leq F_2^-(t)\,|\,X_1 \leq F_1^-(t)),$$

$$\lambda_u := \lambda_{u,12} := \lambda_{u,X_1,X_2} := \lambda_{u,C} := \lim_{t\uparrow 1} \mathbb{P}(X_2 > F_2^-(t)\,|\,X_1 > F_1^-(t)),$$

provided that the limits exist. If $\lambda_l \in (0, 1]$, respectively $\lambda_u \in (0, 1]$, then X_1 and X_2 are *lower tail dependent*, respectively *upper tail dependent*. If $\lambda_l = 0$, respectively $\lambda_u = 0$, then X_1 and X_2 are *lower tail independent*, respectively *upper tail independent*.

Note that there are few known copulas for which the limits in the Definition 1.7.10 do not exist. One such example, a diagonal copula, is constructed in Kortschak and Albrecher (2009) in terms of two other diagonal copulas.

The following result shows that tail dependence is a copula property. It also provides various formulas for computing the tail-dependence coefficients.

Theorem 1.7.11

Let $X_j \sim F_j$, $j \in \{1, 2\}$, be continuously distributed random variables with copula C. Then

(1) $\mathbb{P}(X_2 \leq F_2^-(t)\,|\,X_1 \leq F_1^-(t)) = C(t,t)/t$ for all $t \in (0, 1]$. Thus, λ_l exists if and only if $\lim_{t\downarrow 0} C(t,t)/t$ exists, in which case both are equal.

(2) If $t \mapsto C(t,t)$ is differentiable in a neighborhood of zero and $\lim_{t\downarrow 0} \frac{d}{dt} C(t,t)$ exists, then λ_l exists and equals this limit.

(3) If C is totally differentiable in a neighborhood of zero and $\lim_{t\downarrow 0} (D_1\, C(t,t) + D_2\, C(t,t))$ exists, then λ_l exists and equals this limit.

1 Copulas

Similarly for λ_u.

(4) $\mathbb{P}(X_2 > F_2^-(t) \,|\, X_1 > F_1^-(t)) = (1 - 2t + C(t,t))/(1-t) = \hat{C}(1-t, 1-t)/(1-t)$ for all $t \in [0,1)$. Thus, λ_u exists if and only if $\lim_{t \uparrow 1}(1 - 2t + C(t,t))/(1-t) = \lim_{t \downarrow 0} \hat{C}(t,t)/t$ exists, in which case both are equal.

(5) If $t \mapsto C(t,t)$ is differentiable in a neighborhood of one and $2 - \lim_{t \uparrow 1} \frac{d}{dt} C(t,t)$ exists, then λ_u exists and equals this limit.

(6) If C is totally differentiable in a neighborhood of one and $2 - \lim_{t \uparrow 1}(\mathrm{D}_1 C(t,t) + \mathrm{D}_2 C(t,t))$ exists, then λ_u exists and equals this limit.

Proof

(1) By continuity of F_j, $j \in \{1,2\}$, Proposition 1.1.6 (4), Lemma 1.4.1, and Proposition 1.1.7 (1), it follows that $\mathbb{P}(X_1 \leq F_1^-(t) \,|\, X_2 \leq F_2^-(t)) = \mathbb{P}\bigl(F_1(X_1) \leq F_1(F_1^-(t)), F_2(X_2) \leq F_2(F_2^-(t))\bigr)/\mathbb{P}\bigl(F_1(X_1) \leq F_1(F_1^-(t))\bigr) = \mathbb{P}(F_1(X_1) \leq t, F_2(X_2) \leq t)/\mathbb{P}(F_1(X_1) \leq t) = C(t,t)/t$.

(2) Apply (1) and l'Hôpital's Rule.

(3) Apply (1) and the Chain Rule.

Similarly for (4)–(6). □

1.8 Sampling copulas

Sampling strategies for copulas are interesting from both a theoretical and an empirical perspective. For the former, they often allow for a stochastic representation of the underlying random variables, which may help to access and understand the properties of the underlying dependency. For practical applications, fast sampling algorithms are crucial for large-scale Monte Carlo simulations, e.g., for pricing sophisticated financial products.

The most widely known general sampling algorithm for copulas is the conditional distribution method, presented in this section. To show that this sampling method indeed generates the correct random numbers, let C be a d-dimensional copula and $\boldsymbol{U} \sim C$. Define the *conditional distribution functions of U_j given $U_1 = u_1, \ldots, U_{j-1} = u_{j-1}$* by

$$C(u_j \,|\, u_1, \ldots, u_{j-1}) := \mathbb{P}(U_j \leq u_j \,|\, U_1 = u_1, \ldots, U_{j-1} = u_{j-1}), \; j \in \{2, \ldots, d\}.$$

The following result gives rise to the so-called *conditional distribution method*, see, e.g., Embrechts et al. (2001).

1.8 Sampling copulas

Theorem 1.8.1 (Conditional distribution method)
Let C be a d-dimensional copula, $d \geq 2$, and $\boldsymbol{X} \sim \mathrm{U}[0,1]^d$. Then $\boldsymbol{U} \sim C$, where

$$U_1 := X_1,$$
$$U_2 := C^-(X_2 \,|\, U_1),$$
$$\vdots$$
$$U_d := C^-(X_d \,|\, U_1, \ldots, U_{d-1}).$$

Proof
First consider $d = 2$. By Proposition 1.1.6 (5), $\mathbb{P}(U_1 \leq u_1, U_2 \leq u_2) = \mathbb{P}(X_1 \leq u_1, X_2 \leq C(u_2 \,|\, X_1))$. By Equation (A.1) this equals $\int_0^1 \mathbb{P}(X_1 \leq u_1, X_2 \leq C(u_2 \,|\, X_1) \,|\, X_1 = x_1)\, dx_1 = \int_0^{u_1} C(u_2 \,|\, x_1)\, dx_1 = C(u_1, u_2)$. Now assume the claim holds for some $j - 1 \in \{2, \ldots, d-1\}$. We show that it holds for j. Again by Proposition 1.1.6 (5), $\mathbb{P}(U_1 \leq u_1, \ldots, U_j \leq u_j) = \mathbb{P}(U_1 \leq u_1, \ldots, U_{j-1} \leq u_{j-1}, X_j \leq C(u_j \,|\, U_1, \ldots, U_{j-1}))$. By Identity (A.1) and the induction hypothesis this equals $\int_{[0,1]^{j-1}} \mathbb{P}(u_1' \leq u_1, \ldots, u_{j-1}' \leq u_{j-1}, X_j \leq C(u_j \,|\, u_1', \ldots, u_{j-1}'))\, dC^{(1 \ldots j-1)}(u_1', \ldots, u_{j-1}')$. Since $X_j \sim \mathrm{U}[0,1]$, this equals $\int_{[0,u_1] \times \cdots \times [0,u_{j-1}]} C(u_j \,|\, u_1', \ldots, u_{j-1}')\, dC^{(1 \ldots j-1)}(u_1', \ldots, u_{j-1}')$, thus, by Identity (A.1), $C^{(1 \ldots j)}(u_1, \ldots, u_j)$. □

To find the quantities $C(u_j \,|\, u_1, \ldots, u_{j-1})$, $j \in \{2, \ldots, d\}$, for a specific copula C, the following result is usually applied. It connects the conditional distribution functions with the partial derivatives of C. A rigorous proof can be found in Schmitz (2003, p. 20).

Theorem 1.8.2
Let C be a d-dimensional copula which admits continuous partial derivatives with respect to the first $d-1$ arguments. For $j \in \{2, \ldots, d\}$ and $u_k \in I$, $k \in \{1, \ldots, j\}$,

$$C(u_j \,|\, u_1, \ldots, u_{j-1}) = \frac{\mathrm{D}_{j-1 \ldots 1}\, C^{(1 \ldots j)}(u_1, \ldots, u_j)}{\mathrm{D}_{j-1 \ldots 1}\, C^{(1 \ldots j-1)}(u_1, \ldots, u_{j-1})}.$$

For the case $d = 2$, the assumption of continuity of the partial derivatives in Theorem 1.8.2 can be dropped, in which case the stated equality holds for almost every $u_1 \in I$. If no closed forms for the generalized inverses of the conditional distribution functions $C(u_j \,|\, u_1, \ldots, u_{j-1})$, $j \in \{2, \ldots, d\}$, are available, a numerical root-finding procedure is usually applied to sample random variates from C via Theorem 1.8.1.

1 Copulas

1.9 Elliptical copulas

Among the most famous classes of copulas are Marshall-Olkin copulas, elliptical copulas, and Archimedean copulas. A brief introduction to Marshall-Olkin copulas is given in Section A.2 and Archimedean copulas are presented in Chapter 2 and thereafter. Elliptical copulas arise from elliptical distributions via Sklar's Theorem. In this section, we briefly address elliptical copulas and their basic properties. For more details about elliptical distributions, see, e.g., Cambanis et al. (1981), Fang et al. (1989, pp. 31), and Embrechts et al. (2001).

Definition 1.9.1
Let $\boldsymbol{\mu} \in \mathbb{R}^d$, $\Sigma \in \mathbb{R}^{d \times d}$ a symmetric, positive-semidefinite matrix, and $g : \mathbb{R} \to \mathbb{R}$ some function. The random vector $\boldsymbol{X}$ has an *elliptical distribution* with *location vector* $\boldsymbol{\mu}$, *dispersion matrix* Σ, and *characteristic generator* g, denoted by $\boldsymbol{X} \sim E_d(\boldsymbol{\mu}, \Sigma, g)$, if its characteristic function is given by

$$\phi_{\boldsymbol{X}}(\boldsymbol{t}) := \mathbb{E}[\exp(i\boldsymbol{t}^T \boldsymbol{X})] = \exp(i\boldsymbol{t}^T \boldsymbol{\mu}) g(\boldsymbol{t}^T \Sigma \boldsymbol{t}), \ \boldsymbol{t} \in \mathbb{R}^d. \tag{1.9}$$

If $\boldsymbol{X} \sim E_d(\boldsymbol{\mu}, \Sigma, g) = E_d(\boldsymbol{\mu}^*, \Sigma^*, g^*)$ is *non-degenerate*, i.e., not concentrated on a subset of $\mathbb{R}^d$ of dimension smaller than d, then $\boldsymbol{\mu}^* = \boldsymbol{\mu}$ and there exists some $c > 0$ such that $\Sigma^* = c\Sigma$ and $g^*(t) = g(t/c)$. Thus, for $\boldsymbol{X} \sim E_d(\boldsymbol{\mu}, \Sigma, g)$, Σ and g are only determined up to a positive constant.

If one considers $\boldsymbol{X} \sim E_d(\boldsymbol{0}, I_d, g)$, where I_d denotes the identity matrix in $\mathbb{R}^{d \times d}$, it follows from the definition that $g(\boldsymbol{t}^T \boldsymbol{t})$ must be a characteristic function. As Cambanis et al. (1981) note, this holds if and only if

$$g(t) = \int_{[0,\infty)} h_U(tx^2) \, dF_R(x), \tag{1.10}$$

with h_U such that $h_U(\boldsymbol{t}^T \boldsymbol{t})$ is the characteristic function of $\boldsymbol{U} \sim U(\{\boldsymbol{x} \in \mathbb{R}^k : \boldsymbol{x}^T \boldsymbol{x} = 1\})$ and F_R is a distribution function on $[0, \infty)$. This relation comes into play in the following stochastic representation, where rank Σ denotes the rank of Σ and the equality holds in distribution.

Proposition 1.9.2 (Cambanis et al. (1981))
$\boldsymbol{X} \sim E_d(\boldsymbol{\mu}, \Sigma, g)$ with rank $\Sigma = k$ if and only if

$$\boldsymbol{X} = \boldsymbol{\mu} + R A \boldsymbol{U}, \tag{1.11}$$

where $R \geq 0$, related to g as in (1.10), is independent of the random vector $\boldsymbol{U} \sim U(\{\boldsymbol{x} \in \mathbb{R}^k : \boldsymbol{x}^T \boldsymbol{x} = 1\})$ and $A \in \mathbb{R}^{d \times k}$ is a matrix with rank $A = k$ and $AA^T = \Sigma$.

1.9 Elliptical copulas

From Representation (1.11) it is clear that elliptical distributions are affine transformations of spherical distributions, so they contain the latter as special cases, see McNeil et al. (2005, pp. 90), as well as normal variance mixtures, see McNeil et al. (2005, pp. 73). As Cambanis et al. (1981) note, if $\boldsymbol{X} \sim E_d(\boldsymbol{\mu}, \Sigma, g)$ with rank $\Sigma = d$ and if R is absolutely continuous, then an elliptical distribution admits a density, which has to be of the form $f_{RU}((\boldsymbol{x} - \boldsymbol{\mu})^T \Sigma^{-1}(\boldsymbol{x} - \boldsymbol{\mu}))/\sqrt{\det \Sigma}$, where f_{RU} denotes the existing density of the spherically distributed random vector $R\boldsymbol{U}$. The density of $\boldsymbol{X}$ is thus constant on ellipsoids in $\mathbb{R}^d$, which explains the name of this class of distributions. Elliptical distributions share many useful properties with their most prominent member, the multivariate normal distribution (take $g(t) = \exp(-t/2)$). These properties, summarized in the following proposition, explain the popularity of elliptical distributions and in particular why correlation is a natural measure of association for these distributions.

Proposition 1.9.3

(1) Linearity: If $\boldsymbol{X} \sim E_d(\boldsymbol{\mu}, \Sigma, g)$, $B \in \mathbb{R}^{k \times d}$, and $b \in \mathbb{R}^d$, then $B\boldsymbol{X} + \boldsymbol{b} \sim E_k(B\boldsymbol{\mu} + b, B\Sigma B^T, g)$.

(2) Margins: If $\boldsymbol{X} = (\boldsymbol{X}_1, \boldsymbol{X}_2)^T$ is elliptically distributed with location vector $\boldsymbol{\mu} = (\boldsymbol{\mu}_1, \boldsymbol{\mu}_2)^T$ and dispersion matrix $\Sigma := (\Sigma_{ij})_{i,j \in \{1,2\}}$, then $\boldsymbol{X}_1 \sim E_k(\boldsymbol{\mu}_1, \Sigma_{11}, g)$ and $\boldsymbol{X}_2 \sim E_{d-k}(\boldsymbol{\mu}_2, \Sigma_{22}, g)$, where $k \in \{1, \ldots, d-1\}$ is the dimension of $\boldsymbol{X}_1$.

(3) Conditional distributions: With the notation as before and if Σ is positive definite, then $\boldsymbol{X}_1 | \boldsymbol{X}_2 = \boldsymbol{x}_2 \sim E_k(\boldsymbol{\mu}_1 + \Sigma_{12}\Sigma_{22}^{-1}(\boldsymbol{x}_2 - \boldsymbol{\mu}_2), \Sigma_{11} - \Sigma_{12}\Sigma_{22}^{-1}\Sigma_{21}, g^*)$, see Embrechts et al. (2002) for details about g^*.

(4) Quadratic forms: If $\boldsymbol{X} \sim E_d(\boldsymbol{\mu}, \Sigma, g)$, then $(\boldsymbol{X} - \boldsymbol{\mu})^T \Sigma^-(\boldsymbol{X} - \boldsymbol{\mu}) \sim F_R(\sqrt{\cdot})$, where Σ^- denotes a *generalized inverse* of Σ, i.e., $\Sigma = \Sigma\Sigma^-\Sigma$, and F_R denotes the distribution function of R in Representation (1.11).

(5) Convolutions: If $\boldsymbol{X} \sim E_d(\boldsymbol{\mu}, \Sigma, g)$ and $\boldsymbol{X}^* \sim E_d(\boldsymbol{\mu}^*, c\Sigma, g^*)$, $c \in (0, \infty)$, are independent, then $a\boldsymbol{X} + b\boldsymbol{X}^* \sim E_d(a\boldsymbol{\mu} + b\boldsymbol{\mu}^*, \Sigma, \tilde{g})$ with $\tilde{g}(t) := g(a^2 t) g^*(b^2 ct)$.

By linearity, it follows that for $\boldsymbol{X} \sim E_d(\boldsymbol{\mu}, \Sigma, g)$, $\boldsymbol{X} - \boldsymbol{\mu}$ and $\boldsymbol{\mu} - \boldsymbol{X}$ are equally distributed. Thus, an elliptical distribution is radially symmetric about its location vector.

The multivariate normal distribution is the only elliptical distribution where uncorrelatedness implies independence. Elliptical distributions therefore often serve as counterexamples for this well-known fact.

If $\boldsymbol{X} \sim E_d(\boldsymbol{\mu}, \Sigma, g)$ and $\mathbb{E}[R^2] < \infty$ with R as in Representation (1.11), then $\mathrm{Cov}[\boldsymbol{X}, \boldsymbol{X}] = \Sigma \mathbb{E}[R^2]/d$, see, e.g., Embrechts et al. (2001). Thus, $\rho_{ij} = \Sigma_{ij}/\sqrt{\Sigma_{ii}\Sigma_{jj}}$. If neither X_i nor X_j is concentrated on its mean, Kendall's tau is given by $\tau_{ij} = (1 - \sum_{x \in \mathbb{R}}(\mathbb{P}(X_i = x))^2)(2/\pi) \arcsin \rho_{ij}$,

43

1 Copulas

see Lindskog et al. (2002). A formula for Spearman's rho for elliptical distributions is not known in general, see McNeil et al. (2005, p. 217). However, for the multivariate normal distribution, one has $\rho_{S,ij} = (6/\pi)\arcsin(\rho_{ij}/2)$, see McNeil et al. (2005, p. 215). Concerning tail-dependence coefficients, there is no formula known for elliptical distributions in general. If R is regularly varying, see Hult and Lindskog (2002) for a formula. Since elliptical distributions are radially symmetric, the lower and upper tail-dependence coefficients are necessarily equal. This is considered as one of the major drawbacks of this class of distributions.

One appealing property of Representation (1.11) for elliptical distributions is a straightforward sampling algorithm for $\boldsymbol{X} \sim \mathrm{E}_d(\boldsymbol{\mu}, \Sigma, g)$ applicable in any dimension.

Algorithm 1.9.4 (Elliptical distributions)

(1) Perform a Cholesky decomposition to obtain the Cholesky factor A with $AA^T = \Sigma$.

(2) Sample $\boldsymbol{U} \sim \mathrm{U}(\{\boldsymbol{x} \in \mathbb{R}^d : \boldsymbol{x}^T\boldsymbol{x} = 1\})$.

(3) Sample $R \sim F_R$ independently of $\boldsymbol{U}$.

(4) Return $\boldsymbol{X} := \boldsymbol{\mu} + RA\boldsymbol{U}$.

For Step (1) when Σ is positive definite, see, e.g., Press et al. (2007, pp. 100). For Step (2), see, e.g., McNeil et al. (2005, p. 91), who also show the stochastic representation $\boldsymbol{U} = \boldsymbol{Z}/\|\boldsymbol{Z}\|$ with $\boldsymbol{Z} \sim \mathrm{N}(\boldsymbol{0}, I_d)$.

The copulas arising from elliptical distributions via Sklar's Theorem are called *elliptical copulas*. The most prominent copula family in this class is the *Gaussian copula*, which arises from $\mathrm{E}_d(\boldsymbol{\mu}, \Sigma, \exp(-t/2))$, i.e., the multivariate normal distribution. In d dimensions it takes the form

$$C(\boldsymbol{u}) = \Phi_P(\Phi^{-1}(u_1), \ldots, \Phi^{-1}(u_d))$$
$$= \int_{-\infty}^{\Phi^{-1}(u_d)} \cdots \int_{-\infty}^{\Phi^{-1}(u_1)} \frac{\exp(-\frac{1}{2}\boldsymbol{x}^T P^{-1}\boldsymbol{x})}{\sqrt{(2\pi)^d \det P}} \, dx_1 \ldots dx_d, \quad \boldsymbol{u} \in I^d,$$

where P is the correlation matrix corresponding to Σ, Φ_P the multivariate normal distribution with expectation vector zero and correlation matrix P, and Φ^{-1} the quantile function of the univariate standard normal distribution. Although the copula density is given explicitly by

$$c(\boldsymbol{u}) = \frac{\exp(-\frac{1}{2}\boldsymbol{x}^T(P^{-1} - I_d)\boldsymbol{x})}{\sqrt{\det P}}, \quad \boldsymbol{x} = (\Phi^{-1}(u_1), \ldots, \Phi^{-1}(u_d))^T, \quad \boldsymbol{u} \in I^d,$$

the Gaussian copula itself is not given in explicit form, i.e., it is not an *explicit copula*. For the bivariate case $d = 2$, the entry $\rho = P_{12}$ in the correlation matrix is the correlation coefficient. Concerning the tail-dependence coefficients, it suffices to consider λ_l, which, by Theorem 1.7.11 (3),

1.9 Elliptical copulas

Theorem 1.8.2, and exchangeability equals $\lambda_l = \lim_{t\downarrow 0}(D_1\,C(t,t) + D_2\,C(t,t)) = \lim_{t\downarrow 0}(\mathbb{P}(U_2 \leq t\,|\,U_1 = t) + \mathbb{P}(U_1 \leq t\,|\,U_2 = t)) = 2\lim_{t\downarrow 0}\mathbb{P}(U_2 \leq t\,|\,U_1 = t)$ for $(U_1, U_2)^T$ distributed according to the bivariate Gaussian copula with parameter ρ. The distributional equality $(X_1, X_2)^T = (\Phi^{-1}(U_1), \Phi^{-1}(U_1))^T$ and a simple substitution imply that $\lambda_l = 2\lim_{t\to -\infty}\mathbb{P}(X_2 \leq t\,|\,X_1 = t)$. Since $X_2|X_1 = t \sim \mathrm{N}(\rho t, 1-\rho^2)$, i.e., since X_2 given $X_1 = t$ is normally distributed with mean ρt and variance $1-\rho^2$, it follows that $\lambda_l = 2\lim_{t\to -\infty}\Phi(t(1-\rho)/\sqrt{1-\rho^2}) = 0$ for $\rho < 1$, so the Gaussian copula is not able to capture tail dependence.

For sampling a Gaussian copula, the following steps have to be performed, see McNeil et al. (2005, p. 193).

Algorithm 1.9.5 (Gaussian copula)

(1) Perform a Cholesky decomposition to obtain the Cholesky factor A with $AA^T = P$.

(2) Sample $\mathbf{Z} \sim \mathrm{N}(\mathbf{0}, I_d)$.

(3) Set $\mathbf{X} := A\mathbf{Z}$.

(4) Return $\mathbf{U} := (\Phi(X_1), \ldots, \Phi(X_d))^T$.

The multivariate t_ν distribution with ν degrees of freedom is another well-known elliptical distribution, given by $E_d(\boldsymbol{\mu}, \Sigma, \mathbb{E}[\exp(-Wt/2)])$, where $\nu/W \sim \chi_\nu^2$, $\nu \in (0, \infty)$. Here, χ_ν^2 denotes the chi-square distribution with ν degrees of freedom. Copulas arising from t_ν distributions via Sklar's Theorem are referred to as t_ν *copulas*. They are given by

$$C(\mathbf{u}) = \mathrm{t}_{\nu,P}(\mathrm{t}_\nu^{-1}(u_1), \ldots, \mathrm{t}_\nu^{-1}(u_d))$$

$$= \int_{-\infty}^{\mathrm{t}_\nu^{-1}(u_d)} \cdots \int_{-\infty}^{\mathrm{t}_\nu^{-1}(u_1)} \frac{\Gamma(\frac{\nu+d}{2})}{\Gamma(\frac{\nu}{2})\sqrt{(\pi\nu)^d \det P}}\left(1 + \frac{\mathbf{x}^T P^{-1}\mathbf{x}}{\nu}\right)^{-\frac{\nu+d}{2}} dx_1 \ldots dx_d, \quad \mathbf{u} \in I^d,$$

where P is the correlation matrix corresponding to Σ, $\mathrm{t}_{\nu,P}$ the multivariate t_ν distribution with expectation vector zero and correlation matrix P, and t_ν^{-1} the quantile function of the univariate t_ν distribution. The copula density is given by

$$c(\mathbf{u}) = \frac{\Gamma(\frac{\nu+d}{2})}{\Gamma(\frac{\nu}{2})}\left(\frac{\Gamma(\frac{\nu}{2})}{\Gamma(\frac{\nu+1}{2})}\right)^d \frac{\left(1 + \frac{\mathbf{x}^T P^{-1}\mathbf{x}}{\nu}\right)^{-\frac{\nu+d}{2}}}{\sqrt{\det P}\prod_{j=1}^d \left(1 + \frac{x_j^2}{\nu}\right)^{-\frac{\nu+1}{2}}}, \quad \mathbf{x} = (\mathrm{t}_\nu^{-1}(u_1), \ldots, \mathrm{t}_\nu^{-1}(u_d))^T,$$

$\mathbf{u} \in I^d$. By a similar reasoning as for the Gaussian copula, one can show that the lower and upper tail-dependence coefficients of the bivariate t_ν copula are given by $\lambda_l = \lambda_u = 2\mathrm{t}_{\nu+1}(-\sqrt{\nu+1}\sqrt{1-\rho}/\sqrt{1+\rho})$, where ρ is the off-diagonal element of P, see Demarta and

1 Copulas

McNeil (2004). Thus, t copulas are able to capture tail dependence. However, the lower and upper tail-dependence coefficients are necessarily equal due to radial symmetry. Sampling of t_ν copulas can be achieved with the following algorithm, see McNeil et al. (2005, p. 193).

Algorithm 1.9.6 (t_ν copula)

(1) Sample $\boldsymbol{Z} \sim \mathrm{N}(\boldsymbol{0}, P)$.

(2) Sample $Y \sim \chi_\nu^2$ independently of $\boldsymbol{Z}$ and set $W := \nu/Y$.

(3) Set $\boldsymbol{X} := \sqrt{W}\boldsymbol{Z}$.

(4) Return $\boldsymbol{U} := (\mathrm{t}_\nu(X_1), \ldots, \mathrm{t}_\nu(X_d))^T$.

For more details concerning t distributions and t copulas, we refer the interested reader to the overview of Demarta and McNeil (2004) or the textbook of Kotz and Nadarajah (2004).

2 Archimedean and nested Archimedean copulas

> "Sobald eine Funktion $f(x,y)$ zweier unabhängig veränderlichen Größen x und y die Eigenschaft hat, daß $f(x, f(y,z))$ eine symmetrische Funktion von x, y und z ist, so muß es allemal eine Funktion φ geben, für welche $\varphi(f(x,y)) = \varphi(x) + \varphi(y)$ ist."
>
> Abel (1826)

In contrast to elliptical copulas, Archimedean copulas are not constructed via Sklar's Theorem from known multivariate distributions. Rather, one starts with a given functional form and asks for properties in order to obtain a proper copula. As a result of such a construction, Archimedean copulas are explicit, which is one of the main advantages of this class of copulas. Another advantage of Archimedean copulas is their great flexibility in modeling different kinds of dependencies. Again in contrast to elliptical copulas, Archimedean copulas are not restricted to be radially symmetric. They are therefore able to capture different lower and upper tail dependencies, which is often a desired feature for modeling purposes, e.g., when common losses are observed to happen with a higher probability than common profits. Further, many known copula families are Archimedean, which additionally emphasizes the importance of this class of copulas.

In this chapter, we give an introduction to Archimedean and nested Archimedean copulas. Starting with the former, we briefly cover the bivariate case and then focus on the multivariate setting. In the study of Archimedean copulas in arbitrary dimensions, the Laplace-Stieltjes transform plays a central role. An introduction to this concept is given in Section A.3. In our tour through the Archimedean world, we first focus on conditions under which proper copulas result. The drawbacks of Archimedean copulas then lead us to the more flexible class of nested

2 Archimedean and nested Archimedean copulas

Archimedean copulas. Selected properties of Archimedean and nested Archimedean copulas are discussed afterwards. Finally, examples of known parametric Archimedean families, as well as theoretical sampling algorithms, are presented. As in Chapter 1, for a concise and self-contained introduction, some of the known results are presented with proofs.

2.1 Archimedean copulas

Definition 2.1.1
An *Archimedean generator*, or simply *generator*, is a continuous, decreasing function $\psi : [0, \infty] \to [0, 1]$ which satisfies $\psi(0) = 1$, $\psi(\infty) := \lim_{t \to \infty} \psi(t) = 0$, and which is strictly decreasing on $[0, \inf\{t : \psi(t) = 0\}]$. The set of all such functions is denoted by Ψ. An Archimedean generator $\psi \in \Psi$ is called *strict* if $\psi(t) > 0$ for all $t \in [0, \infty)$. A d-dimensional copula C is called *Archimedean* if it permits the representation

$$C(\boldsymbol{u}) = C(\boldsymbol{u}; \psi) := \psi(\psi^{-1}(u_1) + \cdots + \psi^{-1}(u_d)), \ \boldsymbol{u} \in I^d, \tag{2.1}$$

for some $\psi \in \Psi$ with inverse $\psi^{-1} : [0, 1] \to [0, \infty]$, where $\psi^{-1}(0) := \inf\{t : \psi(t) = 0\}$.

There are different notations for Archimedean copulas circulating. Traditionally, a bivariate Archimedean copula is presented in the form

$$C(\boldsymbol{u}) = \varphi^{[-1]}(\varphi(u_1) + \varphi(u_2)), \ \boldsymbol{u} \in I^2, \tag{2.2}$$

for a continuous, strictly decreasing function $\varphi : [0, 1] \to [0, \infty]$ with $\varphi(0) := \inf\{t : \varphi^{[-1]}(t) = 0\}$ and $\varphi(1) = 0$, where the *pseudo-inverse* $\varphi^{[-1]} : [0, \infty] \to [0, 1]$ is defined by $\varphi^{[-1]}(t) := \varphi^{-1}(t)$ on $[0, \varphi(0))$ and $\varphi^{[-1]}(t) := 0$ otherwise. In this context, φ is the function referred to as Archimedean generator, see, e.g., Schweizer and Sklar (1961), Genest and MacKay (1986), and Nelsen (2007, p. 112). For the general case $d \geq 2$ it turns out to be more convenient to work with Representation (2.1) in terms of the generator ψ. This notation may be found, e.g., in Joe (1997, p. 86) and McNeil and Nešlehová (2009).

Due to the symmetric functional form, a random vector following an Archimedean copula generated by $\psi \in \Psi$ is exchangeable, see Proposition 1.6.2 (2). Finally, let us note that it is clear from Definition 2.1.1 that a generator $\psi \in \Psi$ is only determined uniquely up to a positive constant since $\psi(ct)$ generates the same copula for all $c \in (0, \infty)$.

2.1 Archimedean copulas

2.1.1 Archimedean copulas in two dimensions

Functions of Type (2.2) have a long history. They first appeared long before copulas emerged, in the context of representations for *associative* functions, i.e., functions f that satisfy $f(f(x,y),z) = f(x,f(y,z))$ for all x, y, and z. The early work of Abel (1826) shows that any function $f : \mathbb{R}^2 \to \mathbb{R}$ which is differentiable, strictly monotone in each place, symmetric in its arguments, and associative can be represented as in (2.2). Replacing differentiability by continuity, Aczél (1948) shows that for an open or half-open subinterval J of $\bar{\mathbb{R}}$, any function $f : J^2 \to J$ which is continuous, associative, and strictly increasing in each place can be represented as in (2.2) for a continuous and strictly monotone function φ. Replacing strictly increasing by increasing, Ling (1965) obtains a representation theorem for associative functions which shows that any two-place function $f : [a,b]^2 \to [a,b]$, $a,b \in \bar{\mathbb{R}} : a < b$, which is continuous, associative, increasing in each place, and which satisfies $f(b,x) = x$, $x \in [a,b]$, and $f(x,x) < x$, $x \in (a,b)$, admits the form as given in (2.2) for a continuous and strictly decreasing function $\varphi : [a,b] \to \mathbb{R}$.

Probabilistic metric spaces, introduced by Menger (1942), are generalizations of metric spaces where the metric is replaced by a function mapping two points to a distribution function on the positive real line. The value of this distribution function at x is interpreted as the probability that the distance between the two points is less than x. In such spaces, the notion of a triangle inequality is generalized by the notion of a *triangular norm*, or simply *t-norm*, i.e., a function $T : I^2 \to I$ which is grounded, has uniform margins, is increasing in each place, symmetric in its arguments, and associative. Due to the associativity property, representation theorems for associative functions had an impact on the theory of t-norms. Note that a t-norm is a copula if and only if it is 2-increasing, and a copula is a t-norm if and only if it is associative. Therefore, the theory of t-norms in turn had an influence on copula theory. A first profound result carrying over from t-norms to copulas was a necessary and sufficient condition for φ in order for C as in (2.2) to be indeed a proper copula. Under the assumption of a *strict* t-norm, i.e., a continuous t-norm which is strictly increasing in each place on $(0,1]^2$, this result was obtained by Schweizer and Sklar (1961). A generalization to arbitrary t-norms was given by Schweizer and Sklar (1963), who showed that a function of Type (2.2) with φ and $\varphi^{[-1]}$ as stated above is a t-norm. Another result then shows that a t-norm of Type (2.2) is a copula if and only if φ is convex. Schweizer and Sklar (1963) therefore obtained the following result, see Nelsen (2007, p. 111).

Theorem 2.1.2 (Schweizer and Sklar (1963))
A function C of Type (2.2), with φ and $\varphi^{[-1]}$ as introduced above, is a proper copula if and only if φ is convex.

2 Archimedean and nested Archimedean copulas

Equivalently, a two-place function C of Type (2.1) with $\psi \in \Psi$ is a proper copula if and only if ψ is convex.

The term "Archimedean" entered the scene through the work of Ling (1965) who called a t-norm T *Archimedean* if its *serial iterates* $T^n : I^{n+1} \to I$, defined by $T^2(x_1, x_2) := T(x_1, x_2)$ and $T^{n+1}(x_1, \ldots, x_{n+2}) := T(T^n(x_1, \ldots, x_{n+1}), x_{n+2})$, $n \in \mathbb{N}\setminus\{1\}$, satisfy a version of the *Archimedean property*, namely, for all $x, y \in (0, 1)$ there exists an $n \in \mathbb{N}$ such that $T^n(x, \ldots, x) < y$. In applying the above-mentioned representation theorem for associative functions, Ling (1965) showed that any continuous Archimedean t-norm can be represented as in (2.2), with φ and $\varphi^{[-1]}$ as introduced above. Together with Theorem 2.1.2, this implies that a continuous Archimedean t-norm is a copula if and only if φ is convex. Due to this connection with Archimedean t-norms, copulas of Type (2.2) share the Archimedean property and have thus obtained their name.

With this history in mind, Archimedean copulas are indeed one of the first classes of copulas studied. In fact, in the context of t-norms, Schweizer and Sklar (1961) already present parametric Archimedean generators.

2.1.2 Archimedean copulas in arbitrary dimensions

Similar to the bivariate case, one can ask for necessary and sufficient conditions for an Archimedean generator to generate a proper copula in the multivariate case $d \geq 2$. For answering this question, we need several notions of monotonicity.

Definition 2.1.3

Let $a, b \in \bar{\mathbb{R}} : a < b$ and $f : [a, b] \to \mathbb{R}$ with $f(-\infty) := \lim_{x \to -\infty} f(x)$ if $a = -\infty$ and $f(\infty) := \lim_{x \to \infty} f(x)$ if $b = \infty$, in which case the limits are assumed to exist in the improper sense. Then f is called

(1) *absolutely monotone* on $[a, b]$ if it is continuous there and admits derivatives of all orders satisfying $f^{(k)}(x) \geq 0$ for all $x \in (a, b)$ and $k \in \mathbb{N}_0$;

(2) *d-monotone*, $d \geq 2$, on $[a, b]$ if it is continuous there and admits derivatives up to the order $d - 2$ satisfying $(-1)^k f^{(k)}(x) \geq 0$ for all $k \in \{0, \ldots, d-2\}$, $x \in (a, b)$, and $(-1)^{d-2} f^{(d-2)}(x)$ is decreasing and convex on (a, b);

(3) *completely monotone* on $[a, b]$ if it is continuous there and admits derivatives of all orders satisfying $(-1)^k f^{(k)}(x) \geq 0$ for all $x \in (a, b)$ and $k \in \mathbb{N}_0$.

If the respective monotonicity holds on the whole domain of f, the interval is dropped for simplicity.

2.1 Archimedean copulas

The problem of finding necessary and sufficient conditions on $\psi \in \Psi$ in order to generate a copula for a fixed dimension d has only recently been solved, by McNeil and Nešlehová (2009).

Theorem 2.1.4 (McNeil and Nešlehová (2009))
Let $\psi \in \Psi$ and $d \geq 2$. Then, the function given in (2.1) is a copula if and only if ψ is d-monotone.

In the sequel, we work under a slightly less general setup than in Theorem 2.1.4, but instead have a connection between Archimedean generators and Laplace-Stieltjes transforms at hand. This connection is established in the remaining part of this subsection. We begin with properties of absolutely monotone and completely monotone functions, summarized in the following proposition.

Proposition 2.1.5

(1) The set of completely monotone functions is closed under multiplications and linear combinations with non-negative coefficients, i.e., if f and g are completely monotone, so are fg and $\lambda f + \mu g$ for any $\lambda, \mu \geq 0$.

(2) If f is completely monotone, g non-negative, and g' completely monotone, then $f \circ g$ is completely monotone.

(3) Let f be non-negative such that f' is completely monotone, then $1/f$ is completely monotone.

(4) If f is absolutely monotone and g completely monotone, then $f \circ g$ is completely monotone.

(5) If f is completely monotone, $f(0) = 1$, then f^α is completely monotone for all $\alpha \in (0, \infty)$ if and only if $(-\log f)'$ is completely monotone.

Proof
For (1) and (4), see Widder (1946, p. 145). Part (2) may be found in Feller (1971, p. 441). Part (3) is a straightforward application of (2). For (5), see Joe (1997, p. 374) and Bernstein's Theorem below. □

With the help of generating functions, the following result characterizes absolutely monotone functions on the unit interval. The proof closely follows the one given in Feller (1971, pp. 223).

Lemma 2.1.6
Let $f: [0,1] \to \mathbb{R}$ be continuous. Then the following statements are equivalent:

(1) f is a generating function, i.e., $f(x) = \sum_{k=0}^{\infty} a_k x^k$, $x \in [0,1]$, with $a_k \geq 0$, $k \in \mathbb{N}_0$, and $\sum_{k=0}^{\infty} a_k < \infty$;

(2) f is absolutely monotone on $[0,1]$;

2 Archimedean and nested Archimedean copulas

(3) $\Delta^n_{(x,x+h]}f = \sum_{k=0}^{n}\binom{n}{k}(-1)^{n-k}f(x+kh) \geq 0$ for all $n \in \mathbb{N}$, $x \in (0,1)$, and $h > 0$ with $0 < x + 0h < \cdots < x + nh < 1$.

Proof

To see that (1) implies (2), note that a power series is infinitely-often differentiable in the interior of its disk of convergence. For (2) to (3), let $n \in \mathbb{N}$, $h > 0$, and consider all $x \in (0,1)$ satisfying $0 < x + 0h < \cdots < x + nh < 1$. $f' \geq 0$ implies that f is increasing, which in turn implies that $\Delta_{(x,x+h]}f \geq 0$. Further, $f'' \geq 0$ implies that f' is increasing, thus $(\Delta_{(x,x+h]}f)' = \Delta_{(x,x+h]}f' \geq 0$. Therefore, $\Delta_{(x,x+h]}f$ is increasing, thus $\Delta^2_{(x,x+h]}f \geq 0$. Continuing this way, one obtains $\Delta^n_{(x,x+h]}f \geq 0$. Now consider (3) to (1). Since f is continuous on $[0,1]$, Weierstraß's Theorem implies that the Bernstein polynomials $B_{n,f}(x)$ converge (even uniformly) to $f(x)$ for $n \to \infty$, where $B_{n,f}(x) := \sum_{j=0}^{n} f(j/n)\binom{n}{j}x^j(1-x)^{n-j}$, $n \in \mathbb{N}$. Note that $x^j(1-x)^{n-j} = \sum_{l=0}^{n-j}\binom{n-j}{l}(-1)^l x^{l+j} = \sum_{l=j}^{n}\binom{n-j}{l-j}(-1)^{l-j}x^l$. By changing the order of summation and applying the identity $\binom{n}{j}\binom{n-j}{l-j} = \binom{n}{l}\binom{l}{j}$, this implies that $B_{n,f}(x) = \sum_{l=0}^{n}\binom{n}{l}x^l \sum_{j=0}^{l}\binom{l}{j}(-1)^{l-j}f(j/n)$, thus $B_{n,f}(x) = \sum_{k=0}^{n} a_{nk}x^k$, where, by assumption, $a_{nk} := \binom{n}{k}\Delta^k_{(0,1/n]}f \geq 0$, $k \in \{1,\ldots,n\}$, and $a_{n0} := f(0) \geq 0$. Further, $\sum_{k=0}^{n} a_{nk} = B_{n,f}(1) = f(1) \in [0,\infty)$. Therefore, $\psi_n(t) := B_{n,f}(\exp(-t)) = \mathcal{LS}[F_n](t)$, where F_n is the distribution function of a finite measure which puts mass a_{nk} on $k \in \{0,\ldots,n\}$. With $\psi(t) := f(\exp(-t))$, Theorem A.3.10 (2) implies that $f(\exp(-t)) = \mathcal{LS}[F]$ and that $F_n \to F$ for $n \to \infty$, where F is the distribution function of a finite measure which puts mass on $\mathbb{N}_0$. Thus, f is the generating function corresponding to F. □

Lemma 2.1.6 implies the following result about completely monotone functions, see also Widder (1946, pp. 147).

Lemma 2.1.7

Let $a,b \in \mathbb{R} : a < b$ and $f : [a,b] \to \mathbb{R}$ be continuous. Then f is completely monotone on $[a,b]$ if and only if $\sum_{k=0}^{n}\binom{n}{k}(-1)^{n-k}f(x-kh) \geq 0$ for all $n \in \mathbb{N}$, $x \in (a,b)$, and $h > 0$ with $a < x - nh < \cdots < x - 0h < b$.

Proof

Note that f is completely monotone on $[a,b]$ if and only if $f(b-(b-a)x)$ is absolutely monotone on $[0,1]$. Now apply Lemma 2.1.6 (2), (3) to see that this happens if and only if $\sum_{k=0}^{n}\binom{n}{k}(-1)^{n-k}f(b-(b-a)(x+kh)) \geq 0$ for all $n \in \mathbb{N}$, $x \in (0,1)$, and $h > 0$ with $0 < x + 0h < \cdots < x + nh < 1$. By letting $x' := b - (b-a)x$ and $h' := (b-a)h$, this is equivalent to showing that $\sum_{k=0}^{n}\binom{n}{k}(-1)^{n-k}f(x'-kh') \geq 0$ for all $n \in \mathbb{N}$, $x' \in (a,b)$, and $h' > 0$ with $0 < (b-x')/(b-a) + 0h'/(b-a) < \cdots < (b-x')/(b-a) + nh'/(b-a) < 1$. Multiplying the last inequalities by $-(b-a)$ and adding b leads to the desired result. □

The following result, referred to as *Bernstein's Theorem*, is profound in that it establishes the above-mentioned connection between completely monotone functions on $[0, \infty]$ and Laplace-Stieltjes transforms of distribution functions on $[0, \infty]$, see Feller (1971, pp. 439).

Theorem 2.1.8 (Bernstein (1928))
A function $\psi : [0, \infty] \to [0, 1]$ is the Laplace-Stieltjes transform of a distribution function F on $[0, \infty]$ if and only if ψ is completely monotone on $[0, \infty]$ and $\psi(0) = 1$.

Proof
First consider the "only if" part and note that ψ is continuous on $[0, \infty]$. Since $(-x)^k \exp(-tx)$ is continuous and bounded on $[0, \infty]$ for all $k \in \mathbb{N}_0$ and $t \in (0, \infty)$, we have that $\psi^{(k)}(t) = \int_0^\infty (-x)^k \exp(-tx)\, dF(x)$ for all $k \in \mathbb{N}_0$ and $t \in (0, \infty)$, thus $(-1)^k \psi^{(k)}(t) \geq 0$, $k \in \mathbb{N}_0$, $t \in (0, \infty)$. Further, since F is a distribution function on $[0, \infty]$, $\psi(0) = 1$. For the "if" part, define $h_a(x) := \psi(a(1-x))$, $x \in [0, 1]$, $a \in (0, \infty)$. Since ψ is completely monotone, h_a is absolutely monotone on $[0, 1]$. By Lemma 2.1.6, h_a admits a power series representation. By uniqueness of this representation, the Taylor series expansion of h_a exists, thus, $h_a(x) = \sum_{k=0}^\infty h_a^{(k)}(0)x^k/k! = \sum_{k=0}^\infty (-a)^k \psi^{(k)}(a) x^k/k!$. Now define $\psi_a(t) := \psi\big(a(1 - \exp(-t/a))\big) = h_a(\exp(-t/a)) = \sum_{k=0}^\infty p_k \exp(-kt/a)$, $t \in [0, \infty]$, with $p_k := (-a)^k \psi^{(k)}(a)/k!$, $k \in \mathbb{N}_0$. Then, $p_k \geq 0$, $k \in \mathbb{N}_0$, and $\sum_{k=0}^\infty p_k = h_a(1) = \psi(0) = 1$. Thus, $\psi_a = \mathcal{LS}[F_a]$, where F_a is a distribution function which puts mass p_k on k/a, $k \in \mathbb{N}_0$. By continuity and l'Hôpital's rule, $\lim_{a \to \infty} \psi_a(t) = \psi(t)$, $t \in [0, \infty]$. Since $\psi(0) = 1$, Theorem A.3.10 (2) implies that the limit ψ is the Laplace-Stieltjes transform of a distribution function F on $[0, \infty]$. □

Corollary 2.1.9
Let $\psi \in \Psi$ be completely monotone with $\psi = \mathcal{LS}[F]$. Then $F(0) = 0$ and ψ is strict.

Proof
Assume $F(0) > 0$. Then, by Bernstein's Theorem, $\psi(t) = \int_0^\infty \exp(-tx)\, dF(x) \geq \exp(-t \cdot 0) F(0) > 0$, which contradicts $\psi \in \Psi$. Now, by Bernstein's Theorem and since $F(0) = 0$, we have $\psi(t) = \int_0^\infty \exp(-tx)\, dF(x) \geq \int_0^t \exp(-tx)\, dF(x) \geq \exp(-t^2) \int_0^t dF(x) = \exp(-t^2) F(t) \geq \exp(-t^2)/2$ for all $t \geq t_0$ for t_0 sufficiently large, thus ψ is strict. □

With these tools at hand, we can turn back to the problem of finding necessary and sufficient conditions for an Archimedean generator $\psi \in \Psi$ to generate a proper copula for all dimensions $d \geq 2$. Such conditions are provided by Kimberling (1974) in the context of t-norms. They are already stated in Sklar (1973) in terms of copulas, however, without a proof or reference

2 Archimedean and nested Archimedean copulas

to such. Note that the "if" part of the proof originally presented by Kimberling (1974) can be considerably shortened by using Bernstein's Theorem and the closure of distribution functions under mixtures of products of such.

Theorem 2.1.10 (Kimberling (1974))

Let $\psi \in \Psi$. Then, the function given in (2.1) is a copula for all $d \geq 2$ if and only if ψ is completely monotone.

Proof

For the "only if" part, apply Lemma 2.1.7. Let $n \in \mathbb{N}$, $x \in (0, \infty)$, and $h > 0$ with $0 \leq x - nh < \cdots < x - 0h < \infty$. Note that the case $n = 1$ amounts to showing that $\psi(x - h) - \psi(x) \geq 0$, which holds since $\psi \in \Psi$. Now consider $n \geq 2$. With $a := \psi(\frac{x}{n})$ and $b := \psi(\frac{x}{n} - h)$, it follows from $x - kh = (n-k)\frac{x}{n} + k(\frac{x}{n} - h)$ that $\psi(x - kh) = \psi((n-k)\psi^{-1}(a) + k\psi^{-1}(b))$, thus $\sum_{k=0}^{n} \binom{n}{k}(-1)^{n-k}\psi(x - kh) = \sum_{k=0}^{n}\binom{n}{k}(-1)^{n-k}\psi((n-k)\psi^{-1}(a) + k\psi^{-1}(b))$. This is equal to $\Delta_{(a,b]}C$, where $\boldsymbol{a} := (a, \ldots, a), \boldsymbol{b} := (b, \ldots, b) \in I^n$, $a \leq b$, and thus greater than or equal to zero since, by assumption, C is an n-dimensional copula for all $n \geq 2$. It follows from Lemma 2.1.7 that ψ is completely monotone. For the "if" part, apply Bernstein's Theorem to see that (2.1) can be written as

$$C(\boldsymbol{u}) = \int_0^\infty \prod_{j=1}^d \exp(-x\psi^{-1}(u_j))\, dF(x) = \int_0^\infty \prod_{j=1}^d G(u_j; x)\, dF(x) \qquad (2.3)$$

for $\boldsymbol{u} \in I^d$, where $G(u; x) := \exp(-x\psi^{-1}(u))$, $u \in I$, $x \in (0, \infty)$, and $F = \mathcal{LS}^{-1}[\psi]$. Now note that (2.3) is a mixture of the product of the univariate distribution functions $G(\,\cdot\,; x)$ with respect to the mixture distribution F, thus a multivariate distribution function. Since C has uniform margins, the proof is established. □

Due to the connection with Laplace-Stieltjes transforms, we will mainly consider Archimedean generators ψ belonging to

$$\Psi_\infty := \{\psi \in \Psi : \psi \text{ is completely monotone}\} \subseteq \Psi$$

in the sequel. It will turn out at several points in what follows that the assumption $\psi \in \Psi_\infty$ is rather weak and that generators in Ψ_∞ provide a large subclass of copulas.

2.2 Nested Archimedean copulas

The symmetry of Archimedean copulas is often considered to be a rather strong restriction, especially in large dimensional applications. It implies that all multivariate margins of the same

2.2 Nested Archimedean copulas

dimension are equal, thus, e.g., the dependence among all pairs of components is identical. To allow for asymmetries one may consider the class of nested Archimedean copulas, recursively defined as follows.

Definition 2.2.1

A d-dimensional copula C is called *nested Archimedean* if it is an Archimedean copula with arguments possibly replaced by other nested Archimedean copulas. If C is given recursively by (2.1) for $d = 2$ and

$$C(\boldsymbol{u}; \psi_0, \ldots, \psi_{d-2}) := \psi_0\big(\psi_0^{-1}(u_1) + \psi_0^{-1}(C(u_2, \ldots, u_d; \psi_1, \ldots, \psi_{d-2}))\big), \quad \boldsymbol{u} \in I^d, \qquad (2.4)$$

for $d \geq 3$, C is called *fully-nested Archimedean copula* with $d-1$ *nesting levels* or *hierarchies*. Nesting may also take place in one of the other arguments, not necessarily the last one as done in (2.4). If C is nested Archimedean but not fully-nested Archimedean, then C is called *partially-nested Archimedean copula*.

By definition, nested Archimedean copulas include Archimedean copulas as a special case. The most simple proper fully-nested Archimedean copula C is obtained for $d = 3$ with two nesting levels, given by

$$\begin{aligned} C(\boldsymbol{u}) &= C(u_1, C(u_2, u_3; \psi_1); \psi_0) \\ &= \psi_0\big(\psi_0^{-1}(u_1) + \psi_0^{-1}\big(\psi_1(\psi_1^{-1}(u_2) + \psi_1^{-1}(u_3))\big)\big), \quad \boldsymbol{u} \in I^3. \end{aligned} \qquad (2.5)$$

Its structure can be depicted in form of a tree as in Figure 2.1.

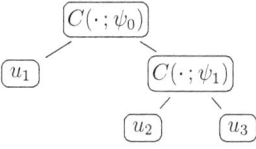

Figure 2.1 Tree structure of a trivariate fully-nested Archimedean copula.

In applications such as the ones discussed in Chapter 5, partially-nested Archimedean copulas

2 Archimedean and nested Archimedean copulas

of the form

$$\begin{aligned}
C(\boldsymbol{u}) &= C(C(u_{11},\ldots,u_{1d_1};\psi_1),\ldots,C(u_{S1},\ldots,u_{Sd_S};\psi_S);\psi_0) \\
&= \psi_0\Big(\psi_0^{-1}\big(\psi_1(\psi_1^{-1}(u_{11})+\cdots+\psi_1^{-1}(u_{1d_1}))\big)+\cdots \\
&\quad +\psi_0^{-1}\big(\psi_S(\psi_S^{-1}(u_{S1})+\cdots+\psi_S^{-1}(u_{Sd_S}))\big)\Big) \\
&= \psi_0\Bigg(\sum_{s=1}^{S}\psi_0^{-1}\bigg(\psi_s\Big(\sum_{j=1}^{d_s}\psi_s^{-1}(u_{sj})\Big)\bigg)\Bigg),\ \boldsymbol{u}\in I^d,
\end{aligned} \qquad (2.6)$$

involving S different *sectors* of dimensions d_s, $s \in \{1,\ldots,S\}$, with $\sum_{s=1}^{S} d_s = d$ arise naturally. The corresponding tree representation is given in Figure 2.2. One can think of the partially-

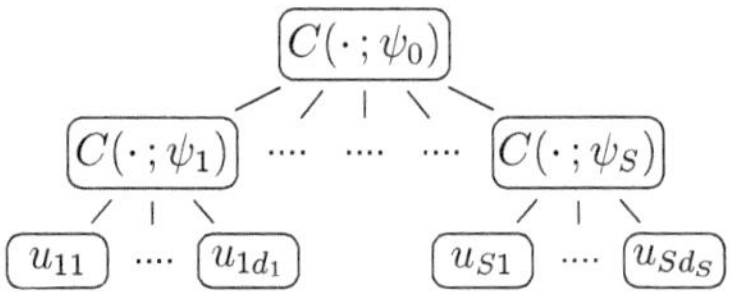

Figure 2.2 Tree structure of a d-dimensional partially-nested Archimedean copula with $\sum_{s=1}^{S} d_s = d$.

nested Archimedean copula of Type (2.6) as capturing the dependence among components in the same sector by some Archimedean copula, and combining these sector copulas by an overall dependence structure, again of the Archimedean type. In other words, all random variables belonging to the same sector s, $s \in \{1,\ldots,S\}$, have the sector copula generated by ψ_s as dependence structure, whereas random variables belonging to different sectors have the copula generated by ψ_0 as common dependence structure.

In the sequel, we will mainly focus on the trivariate fully-nested Archimedean copula of Type (2.5). The arguments readily carry over to the more general cases.

In comparison to Archimedean copulas, there is far less known about nested Archimedean copulas. Although Joe and Hu (1996) investigate dependence structures which generalize the most typical known form of nested Archimedean copulas, the hierarchical functional form and the term "nesting" first appeared in Joe (1997, pp. 87). Since then, this class of copulas appears at several points in the literature, including Whelan (2004), Hofert (2008), McNeil (2008), and Ridout (2008) in the context of sampling algorithms, Savu and Trede (2006) and Okhrin (2007) mainly in the context of statistical inference, and McNeil et al. (2005, pp. 226) in the context of risk management. Other references, specifically in the context of credit risk applications, are

Choroś et al. (2009), Choe and Jang (2009), Höcht and Zagst (2009), and Hofert and Scherer (2010). Due to their hierarchical structure, some of these authors refer to nested Archimedean copulas as *hierarchical Archimedean copulas*.

As in the exchangeable case, one might ask for necessary and sufficient conditions for a nested Archimedean structure to be a proper copula. There is currently only a sufficient condition known, described in what follows. This construction is already mentioned in Joe (1997, p. 88) and implicitly contained in the more general framework of Joe and Hu (1996). A more detailed insight into the argument is given in McNeil (2008). It is probably best described by considering the trivariate case as given in (2.5) and it becomes clear from this case that it applies to the general setting as well. For this, let $\psi_0, \psi_1 \in \Psi_\infty$ with $F_0 := \mathcal{LS}^{-1}[\psi_0]$, see Bernstein's Theorem. For $x \in (0, \infty)$, let $G_0(u; x) := \exp\bigl(-x\psi_0^{-1}(u)\bigr)$, $u \in I$, and

$$\psi_{01}(t; x) := \exp\bigl(-x\psi_0^{-1}(\psi_1(t))\bigr), \ t \in [0, \infty]. \tag{2.7}$$

Further, assume the following *nesting condition* to hold:

$$(\psi_0^{-1} \circ \psi_1)' \text{ completely monotone.} \tag{2.8}$$

By Proposition 2.1.5 (2), this condition implies that $\psi_{01} \in \Psi_\infty$. Note that the trivariate fully-nested Archimedean copula C of Type (2.5) can be written as

$$\begin{aligned} C(\boldsymbol{u}) &= \psi_0\bigl(\psi_0^{-1}(u_1) + \psi_0^{-1}\bigl(\psi_1(\psi_1^{-1}(u_2) + \psi_1^{-1}(u_3))\bigr)\bigr) \\ &= \int_0^\infty \exp(-x\psi_0^{-1}(u_1)) \exp\bigl(-x\psi_0^{-1}\bigl(\psi_1(\psi_1^{-1}(u_2) + \psi_1^{-1}(u_3))\bigr)\bigr) dF_0(x) \\ &= \int_0^\infty G_0(u_1; x)\psi_{01}(\psi_1^{-1}(u_2) + \psi_1^{-1}(u_3); x) \, dF_0(x) \\ &= \int_0^\infty G_0(u_1; x)\psi_{01}\bigl(\psi_{01}^{-1}(G_0(u_2; x); x) + \psi_{01}^{-1}(G_0(u_3; x); x); x\bigr) dF_0(x) \\ &= \int_0^\infty G_0(u_1; x) C(G_0(u_2; x), G_0(u_3; x); \psi_{01}^{-1}(\cdot; x)) \, dF_0(x). \end{aligned} \tag{2.9}$$

From this representation it follows that C is simply an F_0-mixture of the distribution function G_0 times the distribution function constructed via the bivariate Archimedean copula with generator ψ_{01} and equal margins G_0. Thus, C is a distribution function. Since the univariate margins are uniform, C is a copula. The argument for the general case works similarly by deriving the corresponding mixture representation recursively, see McNeil (2008). In essence, for a general nested Archimedean structure to be a proper copula, it is sufficient that all appearing nodes of the form $\psi_i^{-1} \circ \psi_j$ have completely monotone derivatives, i.e., satisfy the nesting condition. For example, for the fully-nested Archimedean copulas of Type (2.4), see the following result of McNeil (2008).

2 Archimedean and nested Archimedean copulas

Theorem 2.2.2 (McNeil (2008))
If $\psi_j \in \Psi_\infty$ for $j \in \{0, \ldots, d-2\}$ such that $\psi_k^{-1} \circ \psi_{k+1}$ have completely monotone derivatives for all $k \in \{0, \ldots, d-3\}$, then $C(\boldsymbol{u}; \psi_0, \ldots, \psi_{d-2})$, $\boldsymbol{u} \in I^d$, is a copula.

Due to the functional forms of the copulas in (2.4) and (2.6), we refer to $F_0 = \mathcal{LS}^{-1}[\psi_0]$ as *outer distribution function*, corresponding to the *outer generator* ψ_0, and to $F_{0s} := \mathcal{LS}^{-1}[\psi_{0s}]$, $s \in \{1, \ldots, S\}$, as *inner distribution functions*, corresponding to the respective *inner generators*. These functions will play an important role in the construction and sampling of nested Archimedean copulas. For notational convenience, we mainly focus on the two functions $F_0 = \mathcal{LS}^{-1}[\psi_0]$ and $F_{01} = \mathcal{LS}^{-1}[\psi_{01}]$ related to the copula in (2.5). More general cases follow directly from the presented ideas, concepts, and strategies.

Remark 2.2.3
It is clear from Representation (2.9) that the nesting condition is sufficient but not necessary. To see this, simply note that any generator ψ_1 which is not completely monotone but such that ψ_{01} generates a proper bivariate Archimedean copula leads to the proper copula representation as given in (2.9), e.g., take $\psi_0(t) := \exp(-t)$, a generator of the independence copula, and $\psi_1(t) := \max\{1 - t, 0\}$, a proper 2-monotone Archimedean generator.

In generalizing this argument, it follows that at least on the lowest level, the nesting condition may be weakened. However, note that the nesting condition is both necessary and sufficient if one requires $\psi_{01} \in \Psi_\infty$. Sufficiency is immediate by Proposition 2.1.5 (1) and necessity is clear by the following argument. It follows from Definition (2.7) that $\psi_{01}(\,\cdot\,; x)$ is completely monotone for all $x \in (0, \infty)$ if and only if $\psi_{01}^\alpha(\,\cdot\,; 1)$ is completely monotone for all $\alpha \in (0, \infty)$, which, by Proposition 2.1.5 (5), is equivalent to the nesting condition.

2.3 Properties of Archimedean and nested Archimedean copulas

The first paper explicitly dealing with Archimedean copulas and their properties is Genest and MacKay (1986). Since then, properties of Archimedean copulas are studied at various points in the literature, including Joe and Hu (1996), Joe (1997, pp. 89), McNeil et al. (2005, pp. 220), and Nelsen (2007, pp. 115). For the most general case of a d-monotone generator, see McNeil and Nešlehová (2009). In that work, the authors discuss precise conditions under which a multivariate Archimedean copula admits a density, level sets, positive lower orthant dependence, and more. They also derive a formula for the distribution function of the *probability integral*

2.3 Properties of Archimedean and nested Archimedean copulas

transformation or *Kendall's transformation*, i.e., $C(\boldsymbol{U})$ for a d-dimensional Archimedean copula C and $\boldsymbol{U} \sim C$, which has been known for d-times continuously differentiable generators since the work of Barbe et al. (1996). In this section we briefly discuss properties of Archimedean and nested Archimedean copulas. We specifically focus on understanding the implied dependence, which is what we need later on, mainly in Chapter 5.

The most appealing property of Archimedean and nested Archimedean copulas is their explicit functional form, which allows one to compute several quantities in closed or semi-closed form. This contrasts elliptical copulas, which do not have a closed form. Another point where these two classes of copulas differ is their symmetry. Elliptical copulas are restricted to be radially symmetric, whereas Archimedean copulas are permutation symmetric in their arguments. The functional symmetry of Archimedean copulas can be relaxed by using nested Archimedean copulas. Nested Archimedean copulas are quite flexible. For example, there are $d!/k!$ possible ways of combining d random variables with the fully-nested Archimedean copula of Type (2.4) and a k-dimensional Archimedean copula on the lowest level, $k \in \{2, \ldots, d\}$. The corresponding copula has $d - k + 1$ parameters and therefore an equal number of possibly different dependencies among the random variables. The number of different partially-nested Archimedean copulas of Type (2.6) are given by the multinomial coefficient $d!/(d_1! \ldots d_S!)$. In general, the number of different ways of combining d random variables with a nested Archimedean copula is given by *Schröder's fourth problem*, see Schröder (1870). For $d = 10$, there are already more than 282 million possible combinations.

The proof of the following result is straightforward and may be found in a slightly different form in Joe (1997, pp. 89) or, for the bivariate case, in Nelsen (2007, p. 136). Note that a function $f : [0, \infty) \to \mathbb{R}$ is *subadditive* if $f(x + y) \leq f(x) + f(y)$ for all $x, y \in [0, \infty)$.

Theorem 2.3.1

Let $\psi_i \in \Psi_\infty$ and C_i be the d-dimensional Archimedean copula generated by ψ_i, $i \in \{0, 1\}$. Then $C_0 \preceq C_1$ if and only if $\psi_0^{-1} \circ \psi_1$ is subadditive.

A sufficient condition for $\psi_0^{-1} \circ \psi_1$ being subadditive is that ψ_0^{-1}/ψ_1^{-1} is increasing on $(0, 1)$, see Nelsen (2007, p. 137). Another one is addressed in the following remark.

Remark 2.3.2

Consider a node of Type (2.7) appearing in a nested Archimedean copula. If the nesting condition holds, then $\psi_0^{-1} \circ \psi_1$ is concave on $[0, \infty)$. Further, $\psi_0^{-1}(\psi_1(0)) = 0$. It follows that $\psi_0^{-1} \circ \psi_1$ is subadditive, see Nelsen (2007, p. 136). Thus, $C_0 \preceq C_1$. By considering $C_0 = \Pi$, using that $-\log \psi_1$ is concave on $[0, \infty)$, see Widder (1946, p. 167), and noting that $-\log(\psi_1(0)) = 0$,

2 Archimedean and nested Archimedean copulas

it follows that $\Pi \preceq C_0$. Thus, under the assumptions of Theorem 2.3.1, we obtain $\Pi \preceq C_0 \preceq C_1$. By Definition 1.7.5 (4), (6), it follows that $0 \leq \kappa_{C_0} \leq \kappa_{C_1}$ for any measure of concordance κ, e.g., $\kappa = \tau$. Translating this to the nodes of the nested Archimedean copula of Type (2.6), this implies that the dependence between two components belonging to the same sector is at least as strong as those between two components belonging to different sectors, measured in terms of concordance. Such dependence structures are often encountered in practice.

Now consider the measures of concordance Spearman's rho and Kendall's tau. For Spearman's rho, there is no explicit formula known for Archimedean copulas in general. However, for a twice continuously differentiable generator ψ with $\psi(t) > 0$, $t \in [0, \infty)$, Kendall's tau can be represented in semi-closed form as

$$\tau = 4 \int_0^1 \frac{\psi^{-1}(t)}{(\psi^{-1}(t))'} \, dt + 1 = 1 - 4 \int_0^\infty t(\psi'(t))^2 \, dt. \tag{2.10}$$

The first equality of Formula (2.10) is established from Identity (1.7) by a transformation, the second equality follows from the first by integration by parts, see Joe (1997, p. 91).

Concerning tail dependence, if the lower and upper tail-dependence coefficients of an Archimedean copula exist and the Archimedean generator is strict, a simple substitution and l'Hôpital's rule lead to the following formulas, see Joe and Hu (1996).

$$\lambda_l = \lim_{t \to \infty} \frac{\psi(2t)}{\psi(t)} = 2 \lim_{t \to \infty} \frac{\psi'(2t)}{\psi'(t)}, \tag{2.11}$$

$$\lambda_u = 2 - \lim_{t \downarrow 0} \frac{1 - \psi(2t)}{1 - \psi(t)} = 2 - 2 \lim_{t \downarrow 0} \frac{\psi'(2t)}{\psi'(t)}. \tag{2.12}$$

Under additional assumptions such as regular variation of ψ or ψ', the tail-dependence coefficients λ_l and λ_u can be derived. Note also the following result.

Proposition 2.3.3

Let $\psi \in \Psi_\infty$ and let λ_l and λ_u denote the tail-dependence coefficients of the Archimedean copula generated by ψ, which are assumed to exist.

(1) If $V \sim F := \mathcal{LS}^{-1}[\psi]$ with $\mathbb{E}[V] < \infty$, then $\lambda_u = 0$.

(2) If $\psi(t) = \sum_{k=1}^\infty p_k \exp(-x_k t)$ for $0 < x_1 < x_2 < \ldots$ and $p_k \geq 0$, $k \in \mathbb{N}$, with $\sum_{k=1}^\infty p_k = 1$, then $\lambda_l = 0$.

Proof

For (1), apply the Frequency Differentiation Theorem, see Section A.3, to see that $-\psi'(0+) =$

$\mathbb{E}[V] < \infty$. This implies

$$\frac{1-\psi(2t)}{1-\psi(t)} = 2\frac{\psi(0)-\psi(2t)}{0-2t}\frac{0-t}{\psi(0)-\psi(t)} \to 2 \ (t \searrow 0).$$

Use the first equality in Formula (2.12) to see that (1) follows. For (2), note that ψ is indeed a member of Ψ_∞. Further, ψ is strict. The result now follows from the first equality in Formula (2.11) since for all $t \in [0, \infty)$,

$$\psi(2t) = \sum_{k=1}^{\infty} p_k \exp(-2x_k t) \leq \exp(-x_1 t) \sum_{k=1}^{\infty} p_k \exp(-x_k t) = \exp(-x_1 t)\psi(t). \qquad \square$$

In contrast to elliptical copulas, Archimedean copulas are not restricted to have equal lower and upper tail-dependence coefficients. As mentioned above, this is one reason for the popularity of Archimedean copulas. As we will see in the following section, the known parametric families of Archimedean copulas constitute a large pool of different tail dependencies.

2.4 Parametric Archimedean families

On the one hand, constructing new Archimedean generators is not difficult, e.g., all generators as addressed in Proposition 2.3.3 (2) are members of Ψ_∞. On the other hand, there are only some parametric Archimedean generators known, which also admit explicit inverses and therefore lead to explicit copulas. Table 2.1 presents known one-parameter Archimedean families with generators $\psi \in \Psi_\infty$. The first column either contains an abbreviation for the name of the respective family, being the one of Ali-Mikhail-Haq ("A"), Clayton ("C"), Frank ("F"), Gumbel ("G"), and Joe ("J"), or the number as given in Nelsen (2007, pp. 116). For simplicity, we choose a slightly different generator for Clayton's family. The second and third column list the corresponding parameter range and parametric generator, respectively, for each Archimedean family. The last column contains either $F = \mathcal{LS}^{-1}[\psi]$ or a stochastic representation for a random variable V following F.

Table 2.1 needs some further explanation. First note that all listed Archimedean generators ψ are elements of Ψ_∞ for all given parameters ϑ. Moreover, they all satisfy $\psi^\alpha \in \Psi_\infty$ for all $\alpha \in (0, \infty)$ and for all parameter choices ϑ. This will be useful at several points in the sequel. For all except for Frank's and Joe's family, this property is seen to hold by applying the tools provided in Proposition 2.1.5. For the families of Frank and Joe, see Joe (1997, p. 375). In Chapter 4 we describe in detail how the listed distributions or stochastic representations are

2 Archimedean and nested Archimedean copulas

Family	ϑ	$\psi(t)$	$V \sim F = \mathcal{LS}^{-1}[\psi]$
A	$[0,1)$	$(1-\vartheta)/(\exp(t)-\vartheta)$	$\text{Geo}(1-\vartheta)$
C	$(0,\infty)$	$(1+t)^{-1/\vartheta}$	$\Gamma(1/\vartheta,1)$
F	$(0,\infty)$	$-\log\bigl(1-(1-e^{-\vartheta})\exp(-t)\bigr)/\vartheta$	$\text{Log}(1-e^{-\vartheta})$
G	$[1,\infty)$	$\exp(-t^{1/\vartheta})$	$S(1/\vartheta,1,\cos^\vartheta(\pi/(2\vartheta)),\mathbb{1}_{\{\vartheta=1\}};1)$
J	$[1,\infty)$	$1-(1-\exp(-t))^{1/\vartheta}$	$\binom{1/\vartheta}{k}(-1)^{k-1},\ k\in\mathbb{N}$
12	$[1,\infty)$	$(1+t^{1/\vartheta})^{-1}$	$S(1/\vartheta,1,\cos^\vartheta(\pi/(2\vartheta)),\mathbb{1}_{\{\vartheta=1\}};1)\,\text{Exp}(1)^\vartheta$
13	$[1,\infty)$	$\exp(1-(1+t)^{1/\vartheta})$	$\tilde{S}(1/\vartheta,1,\cos^\vartheta(\pi/(2\vartheta)),\mathbb{1}_{\{\vartheta=1\}},\mathbb{1}_{\{\vartheta\neq 1\}};1)$
14	$[1,\infty)$	$(1+t^{1/\vartheta})^{-\vartheta}$	$S(1/\vartheta,1,\cos^\vartheta(\pi/(2\vartheta)),\mathbb{1}_{\{\vartheta=1\}};1)\Gamma(\vartheta,1)^\vartheta$
19	$(0,\infty)$	$\vartheta/\log(t+e^\vartheta)$	$\Gamma(\text{Exp}(1)/\vartheta,e^\vartheta)$
20	$(0,\infty)$	$\log^{-1/\vartheta}(t+e)$	$\Gamma(\Gamma(1/\vartheta,1),e)$

Table 2.1 One-parameter Archimedean generators with explicit inverse Laplace-Stieltjes transforms.

found and can be sampled. We therefore omit further details at this point and simply clarify the notation as well as briefly address the basic sampling algorithms involved.

For the family of Ali-Mikhail-Haq, $\text{Geo}(p)$ denotes a *geometric distribution* with success probability $p \in (0,1]$ and mass function $p_k = p(1-p)^{k-1}$ at $k \in \mathbb{N}$. A stochastic representation for a $\text{Geo}(p)$ distributed random variable is given by $\lceil \log(U)/\log(1-p) \rceil$, where $\lceil \cdot \rceil$ denotes the ceiling function and $U \sim \text{U}[0,1]$, see Devroye (1986, pp. 499). For the family of Frank, $\text{Log}(p)$ denotes a *logarithmic distribution* with parameter $p \in (0,1)$ and mass function $p_k = p^k/(-k\log(1-p))$ at $k \in \mathbb{N}$. Here, the sampling algorithm "LK" of Kemp (1981) can be applied. For the family of Joe, $p_k = \binom{1/\vartheta}{k}(-1)^{k-1}$ at $k \in \mathbb{N}$, with $\vartheta \in [1,\infty)$. For sampling discrete distributions with given mass function such as for this Archimedean family, there are several options available, see Devroye (1986, pp. 83). Concerning Clayton's family, $\Gamma(\alpha,\beta)$ denotes the *gamma distribution* with shape $\alpha \in (0,\infty)$, scale $\beta \in (0,\infty)$, and density $f(x) = \beta^\alpha x^{\alpha-1}\exp(-\beta x)/\Gamma(\alpha)$, $x \in (0,\infty)$. For gamma distributions, there is a substantial amount of literature available, see Devroye (1986, pp. 401).

For Gumbel's family, $F = \mathcal{LS}^{-1}[\psi]$ corresponds to a stable distribution. An efficient algorithm for sampling stable distributions is given in Chambers et al. (1976). By definition, a random variable X follows a *stable distribution* if for independent copies X_1 and X_2 of X and all $c_i \in (0,\infty)$, $i \in \{1,2\}$, the random variable $c_1 X_1 + c_2 X_2$ is in distribution equal to $a(c_1,c_2)X + b(c_1,c_2)$ for some constants $a(c_1,c_2) \in (0,\infty)$ and $b(c_1,c_2) \in \mathbb{R}$, see Embrechts et al. (1997, p. 71) or

2.4 Parametric Archimedean families

Nolan (2009, p. 4). There exist various different parameterizations for stable distributions. The most commonly used two are denoted by $S(\alpha, \beta, \gamma, \delta; k)$, $k \in \{0, 1\}$, with characteristic exponent $\alpha \in (0, 2]$, skewness $\beta \in [-1, 1]$, scale $\gamma \in [0, \infty)$, and location $\delta \in \mathbb{R}$. Note that for the two parameterizations $S(\alpha, \beta, \gamma, \delta; k)$, $k \in \{0, 1\}$, the parameters α, β, and γ coincide. For converting the location parameter δ between the two, see Nolan (2009, p. 11). A stable distribution with scale parameter $\gamma = 0$ is understood as the unit jump at δ. Stable distributions are usually best worked with in terms of characteristic functions. The characteristic function of a $S(\alpha, \beta, \gamma, \delta; 0)$ distribution is continuous in all parameters, whereas the one of a $S(\alpha, \beta, \gamma, \delta; 1)$ distribution has a comparably simple functional form, given by

$$\phi(t) = \exp\bigl(i\delta t - \gamma^\alpha |t|^\alpha (1 - i\beta \operatorname{sgn}(t) w(t, \alpha))\bigr), \ t \in \mathbb{R}, \tag{2.13}$$

with

$$w(t, \alpha) = \begin{cases} \tan(\alpha\pi/2), & \alpha \neq 1, \\ -2\log(|t|)/\pi, & \alpha = 1, \end{cases}$$

see Nolan (2009, p. 8). The characteristic function of a $S(1/\vartheta, 1, \cos^\vartheta(\pi/(2\vartheta)), \mathbb{1}_{\{\vartheta=1\}}; 1)$ distribution can be obtained from Equation (2.13) by an application of Euler's formula and it is given by $\phi(t) = \exp(-(-it)^{1/\vartheta})$ for $t \in (0, \infty)$. Thus, the corresponding Laplace-Stieltjes transform equals $\exp(-t^{1/\vartheta})$. Stable distributions appear in the Generalized Central Limit Theorem, see, e.g., Nolan (2009, p. 22), as the precise limiting distributions for properly standardized sums of independent and identically distributed random variables. They are flexible in the sense that they comprise several known distributions, such as the Normal distribution ($\alpha = 2$), the Cauchy distribution ($\alpha = 1$, $\beta = 0$), and the Lévy distribution ($\alpha = 1/2$, $\beta = 1$). This is one of the reasons why stable distributions have been of interest as log-return distributions or distributions of increments of Lévy processes for modeling purposes, see, e.g., Olivares and Seco (2003), Borak et al. (2005), and references therein. The three parameter choices mentioned are the only ones for which the density is given explicitly, whereas in general, it is given by an integral of a rather complicated integrand. By using that $X \sim S(\alpha, \beta, \gamma, \delta; 0)$ if and only if $(X - \delta)/\gamma \sim S(\alpha, \beta, 1, 0; 0)$, see Nolan (2009, p. 9), one can find the general form of the density of a $S(\alpha, \beta, \gamma, \delta; 0)$ distribution by the one of a $S(\alpha, \beta, 1, 0; 0)$ distribution, which is given in Nolan (1997). From this, one can derive the general form of the density of a $S(\alpha, \beta, \gamma, \delta; 1)$ distribution.

The distribution $\tilde{S}(\alpha, \beta, \gamma, \delta, h; 1)$ appearing in Table 2.1 is related to $S(\alpha, \beta, \gamma, \delta; 1)$ and will be addressed in Chapter 4, where we also develop a sampling algorithm for this distribution.

The stochastic representations of V for the families numbered 12 and 14 involve a standard exponentially, respectively a gamma, distributed random variable, which is then multiplied

2 Archimedean and nested Archimedean copulas

by an independently generated stable random variate. The stochastic representations for the families numbered 19 and 20 involve gamma random variables whose parameters themselves depend on independent gamma variables. Further details are given in Chapter 4.

The Archimedean copula families in Table 2.1 that bear a name appear in quite different contexts. The family of Ali-Mikhail-Haq stems from a parametric distribution investigated by Ali et al. (1978) in the context of a certain survival odds ratio. Note that Ali-Mikhail-Haq copulas are precisely the Archimedean copulas with strict generator which are *rational*, i.e., quotients of polynomials, see Nelsen (2007, p. 148). The family of Clayton appears in Clayton (1978). For Frank's family, see Frank (1979) who also shows that these are the only Archimedean copulas which are radially symmetric. For Gumbel's family, see Gumbel (1960). Gumbel copulas are the only Archimedean extreme-value copulas, see, e.g., Nelsen (2007, p. 143). A copula is an *extreme-value copula* if it equals $\lim_{n\to\infty} C^n(\boldsymbol{u}^{1/n})$, $u \in I^d$, for some copula C. The family of Joe appears in Joe (1993).

From a historical point of view it is interesting to note that the Archimedean families of Clayton, Gumbel, and Joe, as well as the family numbered 13, appear in Schweizer and Sklar (1961) in the context of t-norms. As mentioned at the end of Subsection 2.1.1, Archimedean copulas were thus one of the first copulas to be studied.

Table 2.2 gives an overview over the implied dependence. This table contains Kendall's tau, as well as tail-dependence coefficients, for each of the Archimedean families listed in Table 2.1. These entries can be obtained from Formulas (2.10), (2.11), and (2.12). For the families numbered 12, 13, and 14, see Proposition 4.2.8. For Frank's family, $D_1(\vartheta) := \int_0^\vartheta t/(\exp(t)-1)\,dt/\vartheta$ denotes the *Debye function of order one*. For the family numbered 13, $\Gamma(a,b) := \int_b^\infty t^{a-1}\exp(-t)\,dt$ denotes the *upper incomplete gamma function*. For the family numbered 19, $E_1(\vartheta) := \int_\vartheta^\infty \exp(-t)/t\,dt$ denotes the *exponential integral*. For the Archimedean family numbered 20, there is no explicit or at least semi-explicit expression known for Kendall's tau. One may use numerical integration procedures to compute Kendall's tau via Identity (2.10) in this case.

Table 2.3 lists restrictions on the parameters ϑ_i, $i \in \{0,1\}$, such that the corresponding generators ψ_i, $i \in \{0,1\}$, satisfy the nesting condition and thus lead to valid nested Archimedean copulas. With the tools provided in Proposition 2.1.5 and standard results about Laplace-Stieltjes transforms, these conditions are tedious rather than complicated to obtain. The last column of Table 2.3 lists the distributions or stochastic representations of $V_{01} \sim F_{01} = \mathcal{LS}^{-1}[\psi_{01}(\,\cdot\,;V_0)]$. The meaning of the variable V_0 will become clear in Section 2.5. Further, all of these results are derived in Chapter 4. For the moment it suffices to note that for the families of Ali-Mikhail-Haq, Frank, and Joe, V_{01} can be represented as a sum of independent and identically

2.4 Parametric Archimedean families

distributed random variables $V_{01,i}$, $i \in \{1, \ldots, V_0\}$, with given discrete distributions. Thus, F_{01} is a convolution. Since $V_0 \in \mathbb{N}$ for each of these families, the sum is well-defined. For the families numbered 14 and 20, F_{01} is not known.

Family	ϑ	τ	λ_l	λ_u
A	$[0,1)$	$1 - 2(\vartheta + (1-\vartheta)^2 \log(1-\vartheta))/(3\vartheta^2)$	0	0
C	$(0,\infty)$	$\vartheta/(\vartheta+2)$	$2^{-1/\vartheta}$	0
F	$(0,\infty)$	$1 + 4(D_1(\vartheta)-1)/\vartheta$	0	0
G	$[1,\infty)$	$(\vartheta-1)/\vartheta$	0	$2 - 2^{1/\vartheta}$
J	$[1,\infty)$	$1 - 4\sum_{k=1}^{\infty} 1/(k(\vartheta k + 2)(\vartheta(k-1)+2))$	0	$2 - 2^{1/\vartheta}$
12	$[1,\infty)$	$1 - 2/(3\vartheta)$	$2^{-1/\vartheta}$	$2 - 2^{1/\vartheta}$
13	$[1,\infty)$	$1 - (3 - 2^\vartheta e^2 \Gamma(2-\vartheta, 2))/\vartheta$	0	0
14	$[1,\infty)$	$1 - 2/(1 + 2\vartheta)$	$1/2$	$2 - 2^{1/\vartheta}$
19	$(0,\infty)$	$1/3 + 2\vartheta(1 - \vartheta e^\vartheta E_1(\vartheta))/3$	1	0
20	$(0,\infty)$	—	1	0

Table 2.2 Kendall's tau and tail-dependence coefficients for commonly used one-parameter Archimedean generators.

Family	nesting	$V_{01} \sim F_{01} = \mathcal{LS}^{-1}[\psi_{01}(\,\cdot\,;V_0)]$
A	$\vartheta_0 \leq \vartheta_1$	$\sum_{i=1}^{V_0} V_{01,i}$, $V_{01,i} \sim \text{Geo}((1-\vartheta_1)/(1-\vartheta_0))$
C	$\vartheta_0 \leq \vartheta_1$	$\tilde{S}(\vartheta_0/\vartheta_1, 1, (\cos(\pi\vartheta_0/(2\vartheta_1)))V_0)^{\vartheta_1/\vartheta_0}, V_0 \mathbb{1}_{\{\vartheta_0=\vartheta_1\}}, \mathbb{1}_{\{\vartheta_0 \neq \vartheta_1\}}; 1)$
F	$\vartheta_0 \leq \vartheta_1$	$\sum_{i=1}^{V_0} V_{01,i}$, $V_{01,i} \sim \binom{\vartheta_0/\vartheta_1}{k}(-1)^{k-1}(1-e^{-\vartheta_1})^k/(1-e^{-\vartheta_0})$, $k \in \mathbb{N}$
G	$\vartheta_0 \leq \vartheta_1$	$S(\vartheta_0/\vartheta_1, 1, (\cos(\pi\vartheta_0/(2\vartheta_1)))V_0)^{\vartheta_1/\vartheta_0}, \mathbb{1}_{\{\vartheta_0=\vartheta_1\}}; 1)$
J	$\vartheta_0 \leq \vartheta_1$	$\sum_{i=1}^{V_0} V_{01,i}$, $V_{01,i} \sim \binom{\vartheta_0/\vartheta_1}{k}(-1)^{k-1}$, $k \in \mathbb{N}$
12	$\vartheta_0 \leq \vartheta_1$	$S(\vartheta_0/\vartheta_1, 1, (\cos(\pi\vartheta_0/(2\vartheta_1)))V_0)^{\vartheta_1/\vartheta_0}, \mathbb{1}_{\{\vartheta_0=\vartheta_1\}}; 1)$
13	$\vartheta_0 \leq \vartheta_1$	$\tilde{S}(\vartheta_0/\vartheta_1, 1, (\cos(\pi\vartheta_0/(2\vartheta_1)))V_0)^{\vartheta_1/\vartheta_0}, V_0 \mathbb{1}_{\{\vartheta_0=\vartheta_1\}}, \mathbb{1}_{\{\vartheta_0 \neq \vartheta_1\}}; 1)$
14	—	—
19	$\vartheta_0 \leq \vartheta_1$	$\tilde{S}(\vartheta_0/\vartheta_1, 1, (\cos(\pi\vartheta_0/(2\vartheta_1)))V_0)^{\vartheta_1/\vartheta_0}, V_0 \mathbb{1}_{\{\vartheta_0=\vartheta_1\}}, e^{\vartheta_1}\mathbb{1}_{\{\vartheta_0 \neq \vartheta_1\}}; 1)$
20	$\vartheta_0 \leq \vartheta_1$	—

Table 2.3 Nesting conditions and inverse Laplace-Stieltjes transforms corresponding to the inner generator $\psi_{01}(\,\cdot\,;V_0)$.

2 Archimedean and nested Archimedean copulas

All in all, the Archimedean families listed in Table 2.1 provide a rich pool of different dependence structures. In Chapter 4 we will even develop techniques for enlarging this pool. One of these techniques generalizes the notion of inner and outer power families, discussed in the remaining part of this section.

For a generator ψ of a bivariate Archimedean copula, it is clear that $\psi^{1/\alpha}$ also generates a bivariate Archimedean copula for all $\alpha \in (0,1]$. This is due to the fact that $\psi^{1/\alpha}(t)$ is a composition of the increasing, convex function $t^{1/\alpha}$ with the convex function ψ. Similarly, as a composition of the decreasing, convex function ψ with the concave function $t^{1/\beta}$, $\psi(t^{1/\beta})$ also generates a bivariate Archimedean copula for all $\beta \in [1,\infty)$. Families generated by $\psi^{1/\alpha}(t)$ and $\psi(t^{1/\beta})$ are referred to as *inner* (or *interior*) and *outer* (or *exterior*) *power families*, respectively, where the unintuitive use of "inner" and "outer" relates to the fact that they were named with reference to the generator inverses, see Nelsen (2007, p. 142).

A question is to ask if and how these inner and outer power families generalize to the multivariate case. Concerning the former, note that all generators ψ in Table 2.1 satisfy that $\psi^{1/\alpha}$ is completely monotone for any $\alpha \in (0,\infty)$. If our sole knowledge is that ψ is completely monotone, we still know that $\psi^{1/\alpha}$ is completely monotone if $1/\alpha \in \mathbb{N}$, by Proposition 2.1.5 (4). Unfortunately, no general result is known on when we can construct nested Archimedean copulas based on inner power families. Note that the sufficient nesting condition usually does not hold, e.g., consider the generator $\psi(t) = (2e^t - 1)^{-1}$ of Ali-Mikhail-Haq's copula with parameter equal to $1/2$ and let $\psi_i(t) := \psi^{1/\vartheta_i}(t)$ with $\vartheta_i \in (0,1]$, $i \in \{0,1\}$, such that $\vartheta_0/\vartheta_1 = 1/2$. This implies $(\psi_0^{-1} \circ \psi_1)''(\log(5)) > 0$, so $(\psi_0^{-1} \circ \psi_1)'$ is not completely monotone.

In contrast to inner power families, outer power families extend to all dimensions greater than or equal to two, see Proposition 2.1.5 (2). *Outer power copulas*, i.e., copulas based on generators of the form $\tilde{\psi}(t) := \psi(t^{1/\beta})$, $\beta \in [1,\infty)$, provide a way to construct two-parameter Archimedean generators and may therefore add flexibility to the generators given in Table 2.1. In Subsection 4.2.3 we even generalize outer power families and also present some results about the implied dependence. One of these results tells us how the resulting Kendall's tau is related to the Kendall's tau of the copula generated by ψ. For outer power Clayton copulas, e.g., we obtain $\tilde{\tau} = 1 - 2/(\beta(\vartheta + 2))$, where $\vartheta \in (0,\infty)$ denotes the parameter of the Clayton generator ψ. Computing the tail-dependence coefficients from Formulas (2.11) and (2.12) leads to $\tilde{\lambda}_l = 2^{-1/(\beta\vartheta)}$ and $\tilde{\lambda}_u = 2 - 2^{1/\beta}$. These explicit formulas allow us to first choose β such that the desired $\tilde{\lambda}_u \in [0,1)$ is attained, and then choose ϑ such that the desired $\tilde{\lambda}_l \in (0,1)$ is matched. One can also choose one of the tail-dependence coefficients and match a given Kendall's tau. Outer power Clayton copulas are therefore quite flexible. Their nested counterparts will turn out to be particularly useful in Chapter 5.

2.5 Sampling Archimedean and nested Archimedean copulas

This section presents sampling algorithms for Archimedean and nested Archimedean copulas. For this task, there are several options available. One sampling algorithm for Archimedean copulas emerges from a result obtained by Genest and Rivest (1993), in two dimensions, and Barbe et al. (1996), in d dimensions. It is described in Wu et al. (2006) and it extends to the most general case of a d-monotone Archimedean generator, see McNeil and Nešlehová (2009). This algorithm is based on a transformation and requires the sampling of a random variate from the distribution function K of the probability integral transformation of the Archimedean copula. A formula for K for the most general case is given in McNeil and Nešlehová (2009). For a d-dimensional Archimedean copula C, it involves the first $d-2$ derivatives and the left-sided derivative of order $d-1$. Sampling K involves applying the generalized inverse of K to a uniform random variate, see Proposition 1.1.7 (2). This usually requires efficient computation of K and thus knowledge of the derivatives involved. Two sampling algorithms presented by Whelan (2004) are based on a transformation of K but essentially suffer from the same deficiency. One assumes the derivatives to be known explicitly, the other one replaces them by means of the Cauchy Integral Formula but then requires numerical evaluation of the resulting formula, which turns out to be rather slow and may fail in large dimensions, see Whelan (2004).

As a general method, the conditional distribution method presented in Section 1.8 could be tried. This approach for sampling bivariate Archimedean copulas is already presented in Genest and MacKay (1986). Similar to the transformation method mentioned above, applying the conditional distribution method for sampling an Archimedean copula usually requires efficient computation of the generator derivatives, see Theorem 1.8.2. For the Archimedean family of Clayton, one can derive a general formula for the k-th generator derivative. However, accessing the generator or copula derivatives is usually quite demanding. In general, this makes it difficult to apply these algorithms for efficiently sampling Archimedean copulas, especially in large dimensions such as 10, 100, or even larger.

Concerning nested Archimedean copulas, only the conditional distribution method applies in general. For these copulas, however, it seems practically impossible to apply Theorem 1.8.2, even in comparably small dimensions, due to the complicated mixed partial derivatives involved.

These drawbacks bring us to another algorithm for sampling Archimedean copulas, even in large dimensions. Marshall and Olkin (1988) present the following algorithm, referred to as

2 Archimedean and nested Archimedean copulas

Marshall-Olkin algorithm, for sampling an Archimedean copula of Type (2.1). Its correctness is straightforward to show by first conditioning on $V \sim F$ and using that ψ is the Laplace-Stieltjes transform of the distribution function F. As usual, "i.i.d." means "independent and identically distributed".

Algorithm 2.5.1 (Marshall and Olkin (1988))

(1) Sample $V \sim F := \mathcal{LS}^{-1}[\psi]$.

(2) Sample i.i.d. $X_j \sim U[0,1]$, $j \in \{1, \ldots, d\}$, independently of V.

(3) Return $\boldsymbol{U}$, where $U_j := \psi(-\log(X_j)/V)$, $j \in \{1, \ldots, d\}$.

The Marshall-Olkin algorithm is striking. It gives a stochastic representation of the random variables following an Archimedean copula with generator $\psi \in \Psi_\infty$. It is clear from this representation, that these random variables are conditionally independent given V. Being independent given some common random variable is a simple way of introducing dependence. This concept is found in various stochastic models.

As long as the distribution function $F = \mathcal{LS}^{-1}[\psi]$ can be sampled efficiently, the Marshall-Olkin algorithm is fast for all dimensions. This is due to the fact that instead of requiring all the generator derivatives at hand, one only has to know how to sample the inverse Laplace-Stieltjes transform of the Archimedean generator under consideration. Moreover, for families for which this distribution is unknown, there is the option to use procedures for numerically inverting Laplace transforms in order to access the distribution F of V. This is investigated in Chapter 3. A major advantage of the Marshall-Olkin algorithm is that it only requires one random variate V from the univariate distribution F to obtain a vector of random variates from the Archimedean copula, independently of the copula dimension d. The dimension only plays a role for the number of exponential random variates to be drawn. This contrasts the algorithms of Barbe et al. (1996), the sampling algorithms presented by Whelan (2004), and the conditional distribution method, which usually become computationally intense in large dimensions. Note that the idea behind the Marshall-Olkin algorithm generalizes to the case of a d-monotone generator, the Laplace-Stieltjes transform being replaced by the Williamson d-transform, see McNeil and Nešlehová (2009) for more details.

The Marshall-Olkin algorithm may also be derived from Representation (2.3) in the proof of Kimberling's Theorem. One first draws a random variate V from $F = \mathcal{LS}^{-1}[\psi]$. Given V, one then samples the univariate distribution $G(\cdot\,; V)$ according to the desired dimension d. This sampling strategy generalizes to nested Archimedean copulas based on representations such as (2.9). It was first observed and presented with a recursive argument by McNeil (2008). *McNeil's*

2.5 Sampling Archimedean and nested Archimedean copulas

algorithm for sampling a fully-nested Archimedean copula of Type (2.4) is given as follows.

Algorithm 2.5.2 (McNeil (2008))

(1) Sample $V_0 \sim F_0 := \mathcal{LS}^{-1}[\psi_0]$.

(2) Sample $X_1 \sim \mathrm{U}[0,1]$ independently of V_0 and $(X_2, \ldots, X_d)^T \sim C(u_2, \ldots, u_d; \psi_{01}(\cdot\,; V_0), \ldots, \psi_{0d-2}(\cdot\,; V_0))$ independently of X_1.

(3) Return $\boldsymbol{U}$, where $U_j := \psi_0(-\log(X_j)/V_0)$, $j \in \{1, \ldots, d\}$.

Algorithm 2.5.2 is the basic algorithm for sampling nested Archimedean copulas. Note that on the lowest nesting level or recursion step, one always ends up sampling an Archimedean copula which can be achieved with the Marshall-Olkin algorithm.

For sampling partially-nested Archimedean copulas of Type (2.6), McNeil (2008) derived the following algorithm.

Algorithm 2.5.3 (McNeil (2008))

(1) Sample $V_0 \sim F_0 := \mathcal{LS}^{-1}[\psi_0]$.

(2) For $s \in \{1, \ldots, S\}$, sample $(X_{s1}, \ldots, X_{sd_s})^T \sim C(u_{s1}, \ldots, u_{sd_s}; \psi_{0s}(\cdot\,; V_0))$ independently of each other.

(3) Return $\boldsymbol{U} = (U_{11}, \ldots, U_{Sd_S})^T$, where $U_{sj} := \psi_0(-\log(X_{sj})/V_0)$, $s \in \{1, \ldots, S\}$, $j \in \{1, \ldots, d_s\}$.

For sampling the trivariate fully-nested Archimedean copula of Type (2.5), one first samples a random variate V_0 from the distribution F_0 corresponding to ψ_0. Given this random variate one then simply has to sample the bivariate Archimedean copula generated by $\psi_{01}(\cdot\,; V_0)$, see Algorithm 2.5.2 (2). This can be achieved with the Marshall-Olkin algorithm and involves sampling the distribution function F_{01} corresponding to $\psi_{01}(\cdot\,; V_0)$. Efficiently generating the random variates

$$V_0 \sim F_0 = \mathcal{LS}^{-1}[\psi_0] \text{ and } V_{01} \sim F_{01} = \mathcal{LS}^{-1}[\psi_{01}(\cdot\,; V_0)] \quad (2.14)$$

is therefore of major interest and will be investigated in Chapters 3 and 4. As mentioned above, most of the time we will consider the fully-nested Archimedean copula of Type (2.5) as an example. This is only for notational convenience and not a restriction in any way. Due to the recursive character of McNeil's algorithm, the arguments presented for the node $\psi_0^{-1} \circ \psi_1$ readily apply to any node appearing in a nested Archimedean copula. Subsection 4.4.5 contains an example of a fully-nested Archimedean copula involving three nesting levels and Chapter 5 deals

2 Archimedean and nested Archimedean copulas

with partially-nested Archimedean copulas of Type (2.6) with $d = 125$ dimensions involving $S = 6$ sectors on the second level.

3 Numerically sampling nested Archimedean copulas

> *"Most generating functions are sufficiently complicated that the higher derivatives get very complicated quickly and it is not feasible to determine them directly."*
>
> Whelan (2004)

Whether the algorithms of Marshall and Olkin (1988) and McNeil (2008) for sampling Archimedean and nested Archimedean copulas are efficient strongly depends on how well the inverse Laplace-Stieltjes transforms involved in the dependence structure can be accessed. Unfortunately, it is not known how to find these distributions in general. For a given Laplace-Stieltjes transform one might check tables of Laplace transforms, see, e.g., Oberhettinger and Badii (1973). A quite general way to sample these distributions is to use the connection between Laplace-Stieltjes and Laplace transforms via the Integration and Differentiation Theorem in conjunction with Corollary 2.1.9 and to apply algorithms for numerical inversion of Laplace transforms in order to draw the desired random variates. In this chapter, we investigate four algorithms for this task and clarify to what extent these algorithms may be applied for sampling the distributions F_0 and F_{01} as given in (2.14).

Applying an algorithm for numerical inversion of Laplace transforms in order to sample an Archimedean copula is motivated by the fact that, independently of the dimension, one only requires one random variate according to F_0 for constructing a vector of random variates from the copula. For a given Archimedean generator $\psi \in \Psi_\infty$ for which there is no efficient strategy known for sampling $F = \mathcal{LS}^{-1}[\psi]$, it may therefore seem useful to apply a numerical procedure for solving this task. Such a procedure should be sufficiently accurate, since a correctly distributed $V \sim F$ is crucial for building random variates of the corresponding Archimedean copula. Moreover, it should be fast enough to be applicable in large-scale Monte Carlo simulations.

3 Numerically sampling nested Archimedean copulas

We also investigate the precision and run times of the presented algorithms for sampling nested Archimedean copulas. As we will see, sampling nested Archimedean copulas is significantly more difficult than sampling Archimedean copulas.

The results presented in this chapter are published under Hofert (2008). Ridout (2008) also discusses numerical inversion of Laplace transforms for sampling distributions on the positive real line when the distribution function and density are not available. The main algorithms he presents are implemented in the statistical software package R, see R, and a link to the source code can be found in Ridout (2008).

3.1 Numerical inversion of Laplace transforms

The problem of numerically inverting Laplace transforms appears at various points in the literature and plenty of algorithms exist for this task, see Valkó and Vojta (2001). There is no single best algorithm known that works well for all given Laplace transforms, as numerical inversion of Laplace transforms is an *ill-posed* problem, i.e., small changes in values of the Laplace transform can lead to large errors in values of the underlying function. In this section, we present four algorithms for numerically inverting Laplace transforms: the Fixed Talbot, Gaver-Stehfest, Gaver-Wynn-rho, and Laguerre-series algorithm. These algorithms are based on different ideas in order to access the inverse Laplace transform. There are many other algorithms available, see, e.g., Valkó and Vojta (2001) or Cohen (2007), and those we investigate only serve as examples.

3.1.1 The Fixed Talbot algorithm

The Fixed Talbot algorithm is based on the idea of numerically evaluating the Bromwich integral given in Theorem A.3.6 but using a certain choice of the contour to overcome some technical difficulties. In this algorithm, the standard vertical contour $c + ir$, $r \in (-\infty, \infty)$, as given in Theorem A.3.6, is replaced by the contour

$$s_c(t) := ct(\cot(t) + i), \ t \in (-\pi, \pi), \tag{3.1}$$

for a parameter c to the right of all singularities of $\mathcal{L}[f]$. On deforming the standard vertical contour to the contour $s_c(t)$, it is assumed that no singularity of $\mathcal{L}[f]$ is crossed. Applying this change of contour to the Bromwich integral and using $s'_c(t) = ic(1 + i\sigma(t))$ for $\sigma(t) :=$

$t + (t\cot(t) - 1)\cot(t)$ leads to

$$\begin{aligned}f(x) &= \frac{1}{2\pi i}\int_{-\pi}^{\pi}\exp(xs_c(t))\mathcal{L}[f](s_c(t))ic(1+\sigma(t))\,dt \\ &= \frac{c}{2\pi}\int_{-\pi}^{\pi}\operatorname{Re}\Bigl(\exp(xs_c(t))\mathcal{L}[f](s_c(t))(1+\sigma(t))\Bigr)\,dt \\ &= \frac{c}{\pi}\int_{0}^{\pi}\operatorname{Re}\Bigl(\exp(xs_c(t))\mathcal{L}[f](s_c(t))(1+\sigma(t))\Bigr)\,dt\end{aligned}$$

for any continuity point $x \in (0,\infty)$ of f. Computing this integral with the trapezoidal rule with step size π/M leads to the approximation $\tilde{f}$ of f, given by

$$\tilde{f}(x) := \frac{c}{M}\left(\frac{1}{2}\exp(xc)\mathcal{L}[f](c) + \sum_{k=1}^{M-1}\operatorname{Re}\Bigl(\exp(xs_c(t_k))\mathcal{L}[f](s_c(t_k))(1+i\sigma(t_k))\Bigr)\right) \quad (3.2)$$

for $t_k := k\pi/M$, $k \in \{1,\ldots,M-1\}$, where the number of subintervals M is chosen beforehand. The Fixed Talbot algorithm is a further development and simplified version of the algorithm introduced by Talbot (1979). It is explained in Abate and Valkó (2004) who suggest to use $c = 2M/(5x)$ for computing $f(x)$, based on numerical experiments. Because of this "fixed" choice for c, the algorithm is known as the Fixed Talbot algorithm. An implementation of the Fixed Talbot algorithm for computing $\tilde{f}(x)$ as an approximation of $f(x)$ for any continuity point $x \in (0,\infty)$ of f is easily derived from (3.2). In order to control the error of this approximation in an implementation, Abate and Valkó (2004) suggest choosing M as the number of decimal digits that can be represented in the mantissa.

A potential drawback of the Fixed Talbot algorithm is that it requires the evaluation of $\mathcal{L}[f]$ for complex values, but it frequently happens that $\mathcal{L}[f](t)$ is only known for $t \in [0,\infty)$. Note that the extension of $\mathcal{L}[f](t)$ to the complex plane is usually quite simple, as the Laplace transform of f is a holomorphic function on the half-plane of all $s \in \mathbb{C} : \operatorname{Re} s > \operatorname{abs}(\mathcal{L}[f])$. Since it coincides with $\mathcal{L}[f](s)$ (considered as a function with domain $\mathbb{C}$) on the non-negative real line intersected with this half-plane, it suffices to find $a \in \mathbb{R}$ such that $\mathcal{L}[f](s)$ is holomorphic for all $s \in \mathbb{C} : \operatorname{Re} s > a$, as $\mathcal{L}[f](s)$ is then the unique extension of $\mathcal{L}[f](t)$ to $\mathbb{C}$ by the Identity Theorem for holomorphic functions. Therefore, the extension of $\mathcal{L}[f]$ to complex values is usually easy to obtain.

3.1.2 The Gaver-Stehfest and Gaver-Wynn-rho algorithm

The numerical inversion algorithms of the Gaver type are related to another method for recovering f from $\mathcal{L}[f]$, known as Post's Formula, see Theorem A.3.8. If f is continuous and bounded,

3 Numerically sampling nested Archimedean copulas

there is an intuitive derivation for this formula. Fix $x \in (0, \infty)$ and let $(\delta_n(\cdot\,;x))_{n \in \mathbb{N}}$ be a *delta-convergent sequence of densities* on $[0, \infty)$, i.e., the distribution corresponding to δ_n converges, as n tends to infinity, to the point mass at x. An application of the Bounded Convergence Theorem leads to the representation

$$f(x) = \lim_{n \to \infty} \int_0^\infty f(y) \delta_n(y; x) \, dy. \tag{3.3}$$

One choice for $\delta_n(\cdot\,;x)$ is a $\Gamma(n+1, n/x)$ density with expectation $(n+1)x/n$ and variance $(n+1)x^2/n^2$. When computing the right-hand side of (3.3) with this choice for $\delta_n(\cdot\,;x)$, one directly obtains Post's Formula as given in Theorem A.3.8. This formula, however, involves the derivatives of $\mathcal{L}[f]$.

Gaver (1966) carefully chooses

$$\delta_n(y; x) := n \binom{2n}{n} d(x)(1 - \exp(-d(x)y))^n \exp(-nd(x)y) \tag{3.4}$$

$$= n \binom{2n}{n} d(x)((1/2)^{y/x}(1 - (1/2)^{y/x}))^n, \tag{3.5}$$

where $d(x) := \log(2)/x$. Note that (3.4) is the density of the $(n+1)$-th order statistics of a random sample of size $2n$ following an $\text{Exp}(d(x))$ distribution, thus indeed a proper density for all $n \in \mathbb{N}$. It follows from Identity (3.5) and an application of Stirling's formula that $(\delta_n(y; x))_{n \in \mathbb{N}}$ is a delta-convergent sequence. Computing the integral on the right-hand side of (3.3) with δ_n as defined in (3.4), one obtains the so-called *Gaver functionals*, given by

$$f_n(x) := n \binom{2n}{n} d(x) \int_0^\infty f(y)(1 - \exp(-d(x)y))^n \exp(-nd(x)y) \, dy, \; n \in \mathbb{N}.$$

Applying the Binomial Theorem to $(1 - \exp(-d(x)y))^n$, and interchanging integration and summation leads to

$$f_n(x) = n \binom{2n}{n} d(x) \sum_{j=0}^n \binom{n}{j} (-1)^j \mathcal{L}[f]((j+n)d(x))$$

$$= n \binom{2n}{n} (-1)^n d(x) \Delta^n_{[nd(x),(n+1)d(x)]} \mathcal{L}[f], \tag{3.6}$$

where the last equality follows by an index shift. By taking the limit as n tends to infinity, Identity (3.6) looks similar to Post's Formula, but due to the delta-convergent sequence of densities chosen by Gaver (1966), finite differences now replace the derivatives of $\mathcal{L}[f]$. This is advantageous from a computational point of view. As an approximation of f at x, $\tilde{f}(x) = f_M(x)$ for sufficiently large M can be used.

3.1 Numerical inversion of Laplace transforms

Computing the k-th Gaver functional for $k \in \{1, \ldots, n\}$ can be achieved by the following recursion. For $i \in \mathbb{N}_0$ and $j \in \mathbb{N}$, let

$$I_{ij}(x) := \frac{(i+j)!}{i!(j-1)!} d(x) \int_0^\infty f(y)(1 - \exp(-d(x)y))^i \exp(-jd(x)y)\, dy, \quad x \in (0, \infty).$$

It is easily verified that

$$I_{0j}(x) = jd(x)\mathcal{L}[f](jd(x)), \quad j \in \{k, \ldots, 2k\},$$
$$I_{ij}(x) = (1 + j/i)I_{i-1\,j}(x) - (j/i)I_{i-1\,j+1}(x), \quad i \in \{1, \ldots, k\}, \, j \in \{k, \ldots, 2k-i\}.$$

The values I_{ij} can be iteratively computed and stored in an $(n+1) \times 2n$ matrix and the first n Gaver functionals can then be obtained via $f_k(x) = I_{kk}(x)$, $k \in \{1, \ldots, n\}$.

If $f(x)$ admits a Taylor series expansion for all $x \in (0, \infty)$, Gaver (1966) shows that the n-th Gaver functional $f_n(x)$ behaves like $f(x) + \sum_{k=1}^\infty c_k(x)/n^k$ as $n \to \infty$, thus $\lim_{n\to\infty}(f_{n+1}(x) - f(x))/(f_n(x) - f(x)) = 1$, i.e., the Gaver functionals *converge logarithmically* to f. Since this is a rather slow rate of convergence, it is advisable to use a convergence accelerator.

Stehfest (1970) proposes to use a certain linear transformation as convergence accelerator for the Gaver functionals, which means that the n-th member of the transformed sequence is a weighted mean of the first n members of the original sequence. The resulting *Gaver-Stehfest algorithm* utilizes the approximation

$$\tilde{f}(x) := \sum_{k=1}^M w_k f_k(x), \quad x \in (0, \infty),$$

of f, for M sufficiently large, with the *Salzer weights*

$$w_k := (-1)^{M-k} \frac{k^M}{M!} \binom{M}{k}, \quad k \in \{1, \ldots, M\}.$$

The Salzer weights for linear convergence accelerators are discussed in Wimp (1981, pp. 35). Take $\gamma = 1$ in this reference and note the index shift in Wimp (1981, pp. 24). Wimp (1981, pp. 37) shows that these weights provide an *accelerative* linear transformation for a large class of sequences including the logarithmically convergent Gaver functionals. This means that the transformation based on the Salzer weights not only does not change the limit of $(f_k(x))_{k\in\mathbb{N}}$, but the convergence is much faster. The Gaver-Stehfest algorithm is widely used and known to be fast. According to detailed numerical experiments, Valkó and Abate (2004) suggest to work in a computational framework with at least $2M$ decimal digits that can be represented in the mantissa for applying the Gaver-Stehfest algorithm with parameter M.

3 Numerically sampling nested Archimedean copulas

Wynn (1956) suggests the ρ-algorithm, a nonlinear transformation, see Wimp (1981, pp. 168), to accelerate the convergence of sequences. For a given sequence $(y_k)_{k \in \mathbb{N}_0}$, the idea behind this algorithm is as follows. Interpret y_k as the value of a function g at k, $k \in \mathbb{N}_0$. The function g is interpolated with an interpolation formula introduced by Thiele (1909, pp. 132), and the limit for $k \to \infty$ is accessed via extrapolation. To understand this, consider the distinct points $(x_i, y_i)^T$, $i \in \{0, \ldots, n\}$, with $(x_0, y_0)^T := (x, g(x))^T$. Thiele (1909, pp. 129) introduces the *reciprocal differences*, defined by

$$\rho_{i0} := 0, \quad \rho_{i1} := y_i, \quad i \in \{0, \ldots, n\},$$

$$\rho_{ij} := \rho_{i+1\,j-2} + \frac{x_i - x_{i+j-1}}{\rho_{i\,j-1} - \rho_{i+1\,j-1}}, \quad j \in \{2, \ldots, n+1\}, \quad i \in \{0, \ldots, n+1-j\}.$$

Solving the recursively defined term ρ_{ij} with respect to $\rho_{i\,j-1}$ and using $i = 0$ leads to

$$\rho_{0\,j-1} = \rho_{1\,j-1} + \frac{x_0 - x_{j-1}}{\rho_{0j} - \rho_{1\,j-2}}, \quad j \in \{2, \ldots, n+1\}. \tag{3.7}$$

Using $x_0 = x$ and $y_0 = g(x)$, this formula can be recursively applied to find a continued fraction expansion of $g(x)$, given by

$$g(x) = y_1 + \cfrac{x - x_1}{\rho_{12} + \cfrac{x - x_2}{\rho_{13} + \cfrac{x - x_3}{\rho_{14} + \cfrac{x - x_4}{\rho_{05} - \rho_{13}} - \rho_{12}} - y_1} - 0}, \tag{3.8}$$

where the next step would replace ρ_{05} with Formula (3.7) for $j = 6$ and so forth. Stopping this process at some $\rho_{0\,n+1}$ for an even n leads to a rational interpolation function, called *Thiele's interpolation formula*. It interpolates g at the points $(x_i, y_i)^T$ for $i \in \{1, \ldots, n\}$. Further, it is easily verified that $\lim_{x \to \infty} g(x) = \rho_{0\,n+1}$. Thus, $\rho_{0\,n+1}$ yields an approximation of the limit of g for $x \to \infty$, i.e., to the limit of y_k for $k \to \infty$. For example, an approximation of the latter limit for Thiele's interpolation formula as given in (3.8) would be ρ_{05}. Applying the ρ-algorithm to the Gaver functionals $y_k = f_k(x)$ at a fixed $x \in (0, \infty)$ with $x_k = k$ leads to the *Gaver-Wynn-rho algorithm*. According to Valkó and Abate (2004), this accelerator provides the best results among several investigated accelerators. As precision requirement, Abate and Valkó (2004) suggest to use at least $2.1M$ decimal digits that can be represented in the mantissa, according to the chosen parameter M.

3.1.3 The Laguerre-series algorithm

An excellent paper about numerical inversion of Laplace transforms based on Laguerre series is Abate et al. (2006). The main idea is to represent f as a *Laguerre series*, i.e.,

$$f(x) = \sum_{n=0}^{\infty} q_n l_n(x), \; x \in [0, \infty), \tag{3.9}$$

where q_n is the *n-th Laguerre coefficient* and l_n the *n-th Laguerre function*, given by

$$l_n(x) := \exp(-x/2) \sum_{k=0}^{n} \binom{n}{k} \frac{(-x)^k}{k!}, \; x \in [0, \infty), \; n \in \mathbb{N}_0.$$

The Laguerre functions form an orthonormal basis of $L^2([0, \infty))$, see, e.g., Szegö (1975, pp. 100). This implies that Representation (3.9) holds for all $f \in L^2([0, \infty))$ with

$$q_n := \int_0^{\infty} f(y) l_n(y) \, dy, \; n \in \mathbb{N}_0, \tag{3.10}$$

in the L^2-sense. Note that $f \in L^2([0, \infty))$ can often be achieved by *exponentially tilting* f, i.e., considering $\exp(-hx)f(x)$, $h \geq 0$, with corresponding Laplace transform $\mathcal{L}[f](t+h)$.

Applying the Laplace transform to Representation (3.9) leads to $\mathcal{L}[f](t) = \sum_{n=0}^{\infty} q_n \cdot \sum_{k=0}^{n} \binom{n}{k} \cdot I_k(t)/k!$ with $I_k(t) := \int_0^{\infty} (-y)^k \exp(-(t+1/2)y) \, dy$, $k \in \{0, \ldots, n\}$. By using a recursive argument, it follows that $I_k(t) = (-1)^k 2^{k+1} k!/(1+2t)^{k+1}$, $k \in \{0, \ldots, n\}$, thus

$$\mathcal{L}[f](t) = 2 \sum_{n=0}^{\infty} q_n \frac{(2t-1)^n}{(2t+1)^{n+1}}, \; t \in [0, \infty).$$

Now apply the transformation $z := (2t-1)/(2t+1)$ to see that the generating function of the Laguerre coefficients is given by

$$g(z) := \sum_{n=0}^{\infty} q_n z^n = \frac{\mathcal{L}[f]\big((1+z)/(2(1-z))\big)}{1-z} \tag{3.11}$$

for all $z \in D := \{z \in \mathbb{C} : |z| < 1\}$.

Tricomi (1935) and Widder (1935) present the idea of using Equation (3.11) to compute the Laguerre coefficients q_n for $n \in \{0, \ldots, M\}$, where M is some truncation point. Assuming pointwise convergence of the series in (3.9), an approximation $\tilde{f}$ of f can then be computed via

$$\tilde{f}(x) = \sum_{n=0}^{M} q_n l_n(x). \tag{3.12}$$

In order to compute the Laguerre coefficients q_n, $n \in \mathbb{N}_0$, note that $q_0 = \mathcal{L}[f](1/2)$. For computing q_n, $n \in \mathbb{N}$, one can use the *Lattice-Poisson algorithm*, see Abate and Whitt (1992).

3 Numerically sampling nested Archimedean copulas

Szegö (1975, p. 164) showed that $|l_n(y)| \leq 1$ for all $y \in [0, \infty)$. If $f \in L^1([0,\infty))$, Equation (3.10) implies that $|q_n| \leq \int_0^\infty |f(y)|\,dy$, thus the Laguerre coefficients are bounded. It follows that g is finite and analytic on D. An application of the Cauchy Integral Formula then leads to

$$g^{(n)}(z) = \frac{n!}{2\pi i} \oint_{\partial D_r} \frac{g(\zeta)}{(\zeta - z)^{n+1}}\,d\zeta,\ n \in \mathbb{N}_0,\ z \in D_r,$$

where $D_r := \{z \in \mathbb{C} : |z| < r\}$, $r \in (0,1)$, and ∂D_r denotes the boundary of D_r. The choice $z = 0$ and the change of variable $\zeta := r\exp(i\varphi)$ imply

$$q_n = \frac{1}{2\pi i} \oint_{\partial D_r} \frac{g(\zeta)}{\zeta^{n+1}}\,d\zeta = \frac{1}{2\pi r^n} \int_0^{2\pi} g(r\exp(i\varphi))\exp(-ni\varphi)\,d\varphi.$$

Computing this integral with the trapezoidal rule with step size π/n leads to the approximation $\tilde{q}_n$ of q_n, given by

$$\tilde{q}_n := \frac{1}{2nr^n}\Big((-1)^n \operatorname{Re} g(-r) + \operatorname{Re} g(r) + 2\sum_{k=1}^{n-1}(-1)^k \operatorname{Re} g(r\exp(i\pi k/n))\Big),\ n \in \mathbb{N}.$$

Abate and Whitt (1992) derive the corresponding absolute error as $|q_n - \tilde{q}_n| \leq r^{2k}/(1 - r^{2k})$ and propose to choose $r = 10^{-4/n}$ to keep the error bounded by roughly 10^{-8}.

Evaluating the Laguerre functions l_n at $x \in [0, \infty)$ can be done recursively via

$$l_0(x) = \exp(-x/2),\ l_1(x) = (1-x)\exp(-x/2),$$
$$l_n(x) = \frac{2n - 1 - x}{n} l_{n-1}(x) - \frac{n-1}{n} l_{n-2}(x),\ n \geq 2,$$

see Szegö (1975, p. 101).

As a cut-off point for the approximation $\tilde{f}(x)$ of $f(x)$ by (3.12), Abate et al. (2006) propose to use

$$M := \min\{M_0, \min\{n \in \mathbb{N} : \tilde{q}_k < \varepsilon \text{ for all } k \geq n\}\},$$

with $M_0 := 100$ and $\varepsilon := 10^{-8}$. This is the basic *Laguerre-series algorithm*. These authors also discuss problems that might arise from this algorithm and provide several refinements. We only apply the basic algorithm as the suggested refinements take more computational time.

3.2 Sampling Archimedean copulas using numerical inversion algorithms

Inverting Laplace transforms is a standard problem in physics and electrical engineering, where one usually requires high precision rather than fast inversion. For sampling purposes, however,

3.2 Sampling Archimedean copulas using numerical inversion algorithms

we are interested in speed. In this section, we present the Fixed Talbot, Gaver-Stehfest, Gaver-Wynn-rho, and Laguerre-series algorithm for numerically inverting Laplace transforms in order to sample Archimedean and nested Archimedean copulas. We first state the sampling problem to see how such algorithms may be applied. Then, we give specific instructions for sampling a Laplace-Stieltjes inverse involved in an Archimedean or nested Archimedean copula.

Sampling an Archimedean copula with the Marshall-Olkin algorithm requires to draw a random variate from the distribution $F = \mathcal{LS}^{-1}[\psi]$ associated with the Archimedean generator $\psi \in \Psi_\infty$. Applying the Integration and Differentiation Theorem in conjunction with Corollary 2.1.9 leads to the formula

$$\mathcal{L}[F](t) = \frac{\psi(t)}{t}, \ t \in (0, \infty). \tag{3.13}$$

Based on this relation, pointwise evaluation of F can be achieved by applying procedures for numerically inverting Laplace transforms. One can then apply Proposition 1.1.7 (2) for sampling F. For a fixed value $V_0 \in (0, \infty)$, accessing $F_{01} = \mathcal{LS}^{-1}[\psi_{01}(\,\cdot\,; V_0)]$ for sampling nested Archimedean copulas can be achieved in the same way, simply replace ψ by $\psi_{01}(\,\cdot\,; V_0)$ and F by F_{01} in Formula (3.13).

In the remaining part of this section we reformulate the four presented algorithms for pointwise numerically inverting Formula (3.13). The approximation of F at x is denoted by $\tilde{F}(x)$, $x \in [0, \infty)$.

The Fixed Talbot algorithm can be implemented as follows, where we compute as much as possible only once, and as little as required for every value x where we would like to evaluate F.

Algorithm 3.2.1 (Fixed Talbot)

(1) Choose $M \in \mathbb{N}$.

(2) Compute $c_1 := 0.4M$ and $c_2 := e^{c_1}/(2M)$.

(3) For $t_k := k\pi/M$, $k \in \{1, \ldots, M-1\}$, compute $w_{k1} := c_1 t_k (\cot t_k + i)$ and $w_{k2} := e^{w_{k1}}(1 + i(t_k + (t_k \cot t_k - 1) \cot t_k))/w_{k1}$.

(4) Return 0 for $x = 0$ and $\tilde{F}(x) = c_2 \psi(c_1/x) + 0.4 \sum_{k=1}^{M-1} \operatorname{Re}(w_{k2} \psi(w_{k1}/x))$ for $x \in (0, \infty)$.

An implementation of the Gaver-Stehfest algorithm is given as follows. The matrix A is used to store all recursively computed Gaver functionals at x.

Algorithm 3.2.2 (Gaver-Stehfest)

(1) Choose $M \in \mathbb{N}$.

(2) Compute $c := \log(2)$.

3 Numerically sampling nested Archimedean copulas

(3) For $k \in \{1, \ldots, M\}$, compute $w_k := (-1)^{M-k} k^M / (k!(M-k)!)$.

(4) Allocate a matrix $A \in \mathbb{R}^{(M+1) \times 2M}$.

(5) For $x \in (0, \infty)$, compute the $(i, i-1)$-th entry a_{ii-1} of A, $i \in \{1, \ldots, M\}$, via

$$a_{0j} := \psi((j+1)c/x)$$

for $j \in \{0, \ldots, 2M-1\}$ and

$$a_{ij} := (1 + (j+1)/i)a_{i-1j} - ((j+1)/i)a_{i-1j+1}$$

for $i \in \{1, \ldots, M\}$, $j \in \{i-1, \ldots, 2M-i-1\}$.

(6) Return 0 for $x = 0$ and $\tilde{F}(x) = \sum_{k=1}^{M} w_k a_{kk-1}$ for $x \in (0, \infty)$.

The Gaver-Wynn-rho algorithm for solving Equation (3.13) is reworked as follows. Due to the way the matrix B is built, M has to be odd.

Algorithm 3.2.3 (Gaver-Wynn-rho)

(1) Choose $M \in 2\mathbb{N} - 1$.

(2) Compute $c := \log(2)$.

(3) Allocate two matrices $A \in \mathbb{R}^{(M+1) \times 2M}$ and $B \in \mathbb{R}^{M \times (M+1)}$.

(4) For $x \in (0, \infty)$, compute the $(i, i-1)$-th entry a_{ii-1} of A, $i \in \{1, \ldots, M\}$, via

$$a_{0j} := \psi((j+1)c/x)$$

for $j \in \{0, \ldots, 2M-1\}$ and

$$a_{ij} := (1 + (j+1)/i)a_{i-1j} - ((j+1)/i)a_{i-1j+1}$$

for $i \in \{1, \ldots, M\}$, $j \in \{i-1, \ldots, 2M-i-1\}$.

(5) Compute the $(0, M)$-th entry b_{0M} of B via

$$b_{i0} := 0, \ b_{i1} := a_{i+1i}, \ i \in \{0, \ldots, M-1\},$$
$$b_{ij} := b_{i+1j-2} + \frac{j-1}{b_{i+1j-1} - b_{ij-1}}, \ j \in \{2, \ldots, M\}, \ i \in \{0, \ldots, M-j\}.$$

(6) Return 0 for $x = 0$ and $\tilde{F}(x) = b_{0M}$ for $x \in (0, \infty)$.

Finally, consider the Laguerre-series algorithm. Since F is neither in $L^1([0,\infty))$ nor in $L^2([0,\infty))$, we use the relation $\mathcal{L}[\exp(-x)F(x)](t) = \mathcal{L}[F](t+1) = \psi(t+1)/(t+1)$, $t \in (0,\infty)$, instead of Equation (3.13), to access F at x via $\exp(-x)F(x)$. The Laguerre-series algorithm for accessing F can then be formulated as follows.

Algorithm 3.2.4 (Laguerre-series)

(1) Choose $M_0 \in \mathbb{N}$ and $\varepsilon > 0$.

(2) Compute $q_0 := 2\psi(3/2)/3$.

(3) Set $k = 0$. Increase k by one and compute $\tilde{q}_k$ via

$$\tilde{q}_k := \frac{1}{2kr^k}\left((-1)^k \operatorname{Re} g(-r) + \operatorname{Re} g(r) + 2\sum_{j=1}^{k-1}(-1)^j \operatorname{Re} g(re^{i\pi j/k})\right)$$

with $g(z) := 2\psi(3-z)/(3-z)$ and $r = 10^{-4/k}$ until $\tilde{q}_k < \varepsilon$ or $k \geq M_0$.

(4) Set $M := k$.

(5) For $x \in (0,\infty)$, compute $l_0(x) := \exp(-x/2)$ and $l_1(x) := (1-x)\exp(-x/2)$. Compute the k-th Laguerre function $l_k(x)$, $k \in \{2,\ldots,M\}$, via

$$l_k(x) := \frac{2k-1-x}{k}l_{k-1}(x) - \frac{k-1}{k}l_{k-2}(x).$$

(6) Return 0 for $x = 0$ and $\tilde{F}(x) = \exp(x)\sum_{k=0}^{M}\tilde{q}_k l_k(x)$ for $x \in (0,\infty)$.

3.3 Experiments and examples

In this section, we present and discuss several experiments and give examples of how Archimedean and nested Archimedean copulas can be sampled with algorithms for numerically inverting Laplace transforms. We start with a precision and run-time comparison for the four different approaches for sampling the Archimedean family of Clayton. Afterwards, we apply the direct approach, the conditional distribution method, and the Fixed Talbot algorithm to sample a Clayton copula. We then sample an outer power Archimedean copula based on Clayton's generator. Since the inverse Laplace-Stieltjes transform of the outer power Clayton generator is not known, we apply the Fixed Talbot algorithm to access it. A nested version of this copula then follows. For this, the inverse Laplace-Stieltjes transform corresponding to the outer distribution function is not known, whereas the distribution function corresponding to the inner Archimedean generator is known. Afterwards, we discuss the use of numerical inversion of Laplace transforms

3 Numerically sampling nested Archimedean copulas

for sampling nested Gumbel and Clayton copulas. As a last example, we consider a nested Archimedean copula based on a Clayton copula nested into an Ali-Mikhail-Haq copula. For this example, the outer but not the inner distribution function is known explicitly. For illustrative purposes, the nested Archimedean copulas only involve two nesting levels. However, note that due to the recursive character of McNeil's algorithm, more complex hierarchies could be sampled equally well, see Subsection 4.4.5.

3.3.1 A word concerning the implementation

In the remaining part of this chapter, as well as in the following chapters, we frequently present and discuss results obtained by simulation studies. In order for these studies to be comparable, one has to provide the precise setup under which the results were obtained. This subsection is devoted to this task.

All algorithms are implemented in C/C++ and compiled using the GCC 3.3.3 (SuSE Linux) with option -O2 for code optimization. The computer programs are run on a node containing two AMD Opteron 252 processors with 2.6 GHz and 8 GB RAM as part of the Linux cluster "Pacioli", see Pacioli. We would hereby like to thank the Scientific Computing Centre Ulm, see UZWR, for providing access to this cluster. Under double precision, the number of base-10 digits that can be represented in the mantissa is 15, reported by the function digits10 of the C/C++ library limits. The smallest strictly positive floating-point number in double precision is $x_{\min} := 2.2251 \cdot 10^{-308}$, given by DBL_MIN in cfloat. The command gettimeofday is used to measure run time as wall-clock time, denoted by κ. Apart from the physical structure of the grid, the compiler, and the programming language, note that run time also depends on other factors such as the quality of the implementation or the current load of the machine. The presented run times should therefore be viewed with this in mind. For generating uniform random variates an implementation of the Mersenne Twister by Wagner (2003) is used. For more details concerning this pseudorandom number generator, see Matsumoto and Nishimura (1998). For numerical root finding, we use the function nag_zero_cont_func_bd_1 of the NAG library, see NAG. The corresponding parameters xtol and ftol are chosen as 10^{-8} and 0, respectively, except for the outer power Clayton copula in Subsection 3.3.4, where ftol is chosen as 10^{-8} in order to have convergence in root finding for all drawn variates.

In the presented examples, we often need to find the U-quantile of a distribution function for a realization $U \sim \mathrm{U}[0,1]$ close to one. If an algorithm is not able to locate the U-quantile of F as the computed values $F(x)$ for reasonably large x are smaller than U, we use *truncation*,

3.3 Experiments and examples

i.e., we specify a maximal value $x_{\max}$ where F is evaluated and return $x_{\max}$ if $U > F(x_{\max})$. If $F(x_{\max}) \geq 1 - \varepsilon$ for some $\varepsilon \in (0, 1)$, at most 1 per $1/\varepsilon$ drawn random variates of F is expected to be obtained by truncation and therefore causes a bias in sampling F. In Chapter 4, we will also consider discrete distribution functions F with jumps at $\mathbb{N}$. Their values at $\{1, \ldots, x_{\max}\}$, $x_{\max} \in \mathbb{N}$, can be computed and stored. If $U > F(x_{\max})$, we return $x_{\max} + 1$, otherwise, we locate the U-quantile in the table of precomputed values. Such *table look-up methods* seem useful in this setup, especially if a large amount of random variates has to be sampled. Note that we can even avoid errors due to truncation by using the following argument. Assume we want to sample n variates. One strategy for this is to first draw all n required random variates $U \sim \mathrm{U}[0, 1]$, determine the maximum $U_{(n)}$ of these variates, and then compute and store the values of F up to $U_{(n)}$ in a table. This way, no truncation takes place. Besides truncation, other methods may be applied. One involves Tauberian theorems to obtain asymptotics in the tail, see, e.g., Feller (1971, pp. 442).

For verifying the correctness of the generated vectors of random variates, our studies include the trivariate case. We implement several quality checks of the generated data. For the first check, each dimension is subdivided into five equally spaced bins. The corresponding grid partitions the three-dimensional unit interval into 125 cubes. For every trivariate example, 1 000 sets of samples, each of size 100 000, are taken and the p-value of the Kolmogorov-Smirnov test based on the 1 000 computed χ^2 test statistics is reported. Unless otherwise stated, the expected number of variates in each bin is greater than or equal to five, being a rule of thumb for the χ^2 test. The second check consists of computing pairwise sample versions of Kendall's tau $\hat{\tau}_{ij}$ for $i, j \in \{1, 2, 3\}$ with $i < j$ based on a set of 100 000 three-dimensional vectors of random variates. We only consider these quality checks in the trivariate case, since they get computationally intense in larger dimensions. It is intuitively clear that deviations from the intended dependence structure should also be detectable in the trivariate case. Basic quality checks such as testing the variates for being floating-point numbers in the unit interval are conducted in all dimensions. Note that plots may serve as another quality check for detecting major flaws of numerical inversion algorithms. Whenever we sample an Archimedean copula with generator $\psi = \mathcal{LS}[F]$, respectively a nested Archimedean copula involving an outer generator $\psi_0 = \mathcal{LS}[F_0]$ and inner generators of the form $\psi_{01} = \mathcal{LS}[F_{01}]$, we plot the approximation $\tilde{F}(x)$, respectively $\tilde{F}_0(x)$ and $\tilde{F}_{01}(x)$, for all x on an equally spaced, fine grid. For plotting $\tilde{F}_{01}$, reasonable realizations V_0 of F_0 are chosen.

3 Numerically sampling nested Archimedean copulas

3.3.2 Precision and run-time comparison for Clayton's copula

One of the first papers to study and compare the performance, regarding precision and run time, of different algorithms for inverting Laplace transforms is Davies and Martin (1979). In this subsection, we compare the Fixed Talbot, Gaver-Stehfest, Gaver-Wynn-rho, and Laguerre-series algorithm regarding precision and run time for evaluating F for a Clayton copula. For this Archimedean copula, F is easily verified to be a $\Gamma(1/\vartheta, 1)$ distribution, thus one can compare the approximate values $\tilde{F}(x)$ with their theoretical counterparts $F(x)$ for $x \in [0, \infty)$. As parameter we choose $\vartheta = 0.8$, corresponding to a Kendall's tau of 0.2857.

In order to compare the precision of the different algorithms, we evaluate the gamma distribution function F at 10 000 equally spaced points, ranging from 0.001 to 10. The set of these points is denoted by P. As $F(0.001)$ is close to zero and $F(10)$ is close to one, these values should give an insight into the overall approximation of F by $\tilde{F}$. Similar to Abate and Valkó (2004), we measure the precision of this approximation with the *maximal relative error*, given by

$$\text{MXRE} := \max_{x \in P} \left| \frac{F(x) - \tilde{F}(x)}{F(x)} \right|$$

and consider a procedure for numerically inverting Laplace transforms to be accurate enough for our sampling problem as given in (3.13) if MXRE satisfies

$$\text{MXRE} < 0.0001. \tag{3.14}$$

We also report the *mean relative error*, defined by

$$\text{MRE} := \frac{1}{|P|} \sum_{x \in P} \left| \frac{F(x) - \tilde{F}(x)}{F(x)} \right|.$$

Table 3.1 presents the results of our investigation for all four algorithms. The second column lists the choices of the parameter M for the Fixed Talbot, Gaver-Stehfest, and Gaver-Wynn-rho algorithm. Note that the precision requirements of Abate and Valkó (2004) hold for the given choices of M. For the Laguerre-series algorithm, the second column of Table 3.1 addresses different choices of the parameter ε. Note that we do not use the bound M_0 in order to obtain a better insight into the precision of this algorithm. Table 3.1 also reports if numerical difficulties were observed, i.e., if values $\tilde{F}(x)$ which are not in the unit interval for at least one $x \in P$ are obtained. Finally, MXRE and MRE, as well as the run times κ in seconds for 1 000 evaluations of $\tilde{F}(x)$ for all points $x \in P$, are listed.

According to the results reported in Table 3.1, the Laguerre-series algorithm can not compete with the other algorithms for solving Equation (3.13) when F is a $\Gamma(5/4, 1)$ distribution. For

3.3 Experiments and examples

Algorithm	parameter	difficulties	MXRE	MRE	κ
Fixed Talbot	1	no	1.64990123	0.30362935	2.52
	2	yes	0.07726105	0.01880226	5.50
	3	no	0.01039537	0.00384815	9.21
	4	yes	0.00228519	0.00104870	12.43
	5	no	0.00050258	0.00019882	16.14
	6	no	0.00008386	0.00002621	19.77
	7	no	0.00001308	0.00000516	23.49
	8	no	0.00000280	0.00000119	27.15
	9	no	0.00000064	0.00000010	31.10
	10	no	0.00000008	0.00000004	34.64
	11	no	0.00000005	0.00000001	38.64
	12	no	0.00000005	0.00000000	41.95
	13	no	0.00000005	0.00000000	45.19
	14	no	0.00000005	0.00000000	49.39
	15	no	0.00000005	0.00000000	53.07
Gaver-Stehfest	1	no	1.82544975	0.11408466	5.03
	2	no	0.14920303	0.02117223	10.07
	3	yes	0.01475319	0.00479624	15.16
	4	yes	0.00309887	0.00145811	20.59
	5	yes	0.00082358	0.00047443	25.94
	6	no	0.00027974	0.00012669	31.56
	7	no	0.00008557	0.00003468	37.24
Gaver-Wynn-rho	1	no	1.82544975	0.11408466	5.03
	3	yes	53.52189421	0.04938593	15.42
	5	yes	0.00047227	0.00020851	27.02
	7	no	0.00008025	0.00000831	39.21
Laguerre-series	10^{-6}	yes	0.62016824	0.00072673	35.07
	10^{-7}	yes	0.02712199	0.00008886	95.97
	10^{-8}	yes	0.02291828	0.00001404	265.95
	10^{-9}	no	0.00095077	0.00000158	743.13
	10^{-10}	no	0.00015965	0.00000022	2100.40

Table 3.1 Precision and run times in seconds for 1 000 evaluations of $\tilde{F}(x)$ for all $x \in P$.

3 Numerically sampling nested Archimedean copulas

$\varepsilon = 10^{-9}$, 5453 and for $\varepsilon = 10^{-10}$, 15164 Laguerre coefficients were computed. For $\varepsilon = 10^{-10}$, one almost reaches an acceptable precision, but the evaluation takes an unacceptable amount of time. Although the Laguerre coefficients are only computed once for the 1000 evaluations of the 10000 points in P, the run time is large due to the number of Laguerre functions that have to be computed. We therefore did not further consider the Laguerre-series algorithm for sampling purposes. However, note that the performance of this algorithm may be significantly better for other distributions F, maybe also in conjunction with certain sequence accelerators.

According to the precision requirements recommended for the Gaver-Stehfest and the Gaver-Wynn-rho algorithm by Abate and Valkó (2004), `double` precision should only be used for the parameter values of $M \in \{1, \ldots, 7\}$, otherwise, a multi precision framework is recommended. However, working under such a framework usually increases run time enormously, which we think is not appropriate for sampling large amounts of random numbers. We therefore consider only `double` precision and investigate different choices of M there. The Gaver-Stehfest and Gaver-Wynn-rho algorithm perform well for $M = 7$ and meet Requirement (3.14). The run times for this choice of M are also quite impressive for the 10 million function evaluations. Moreover, these algorithms only work with real arguments of the underlying generator.

In this precision and run-time comparison, the Fixed Talbot algorithm performed the best. Both the maximal relative error MXRE and the mean relative error MRE are decreasing in M, the latter even vanishes according to the first eight decimal places. Already for the parameter choice $M = 6$, this algorithm satisfies Requirement (3.14). Moreover, it shows high speed at the same time.

Although one certainly can not give a precise recommendation which algorithm to use for solving Equation (3.13) for general F, the results of Table 3.1 indicate that procedures for numerically inverting Laplace transforms are indeed an option to achieve this task. For inverting the chosen gamma distribution, the Fixed-Talbot algorithm outperformed the Gaver-Stehfest, Gaver-Wynn-rho, and Laguerre-series algorithm regarding precision and run time. The Gaver-Stehfest and Gaver-Wynn-rho algorithm also performed well. Thus, there are adequate alternatives if the Fixed-Talbot algorithm might have difficulties in accessing a particular F.

3.3.3 A Clayton copula

In our first example, we consider a Clayton copula, with generator $\psi(t) = (1+t)^{-1/\vartheta}$ corresponding to a $\Gamma(1/\vartheta, 1)$ distribution. The Clayton generator ψ is the only known Archimedean generator for which all derivatives can easily be derived. This example is therefore suitable for a

3.3 Experiments and examples

comparison of the performance of directly sampling $V \sim \Gamma(1/\vartheta, 1)$, the Fixed Talbot algorithm, and the conditional distribution method.

Note that for an Archimedean copula with generator $\psi \in \Psi_\infty$, the conditional distribution functions $C(u_j \mid u_1, \ldots, u_{j-1})$ as appearing in Theorem 1.8.2 can be obtained by

$$C(u_j \mid u_1, \ldots, u_{j-1}) = \frac{\psi^{(j-1)}\left(\sum_{k=1}^{j} \psi^{-1}(u_k)\right)}{\psi^{(j-1)}\left(\sum_{k=1}^{j-1} \psi^{-1}(u_k)\right)}, \; j \in \{2, \ldots, d\}. \tag{3.15}$$

For Clayton's family, $\psi^{(k)}(t) = (-1)^k (1+t)^{-(k+1/\vartheta)} \prod_{i=0}^{k-1}(i+1/\vartheta)$, $k \in \mathbb{N}$. Thus, we may conclude that

$$C^{-}(x \mid u_1, \ldots, u_{j-1}) = \psi\left(\left(1 + \sum_{k=1}^{j-1} \psi^{-1}(u_k)\right)(x^{-1/(j-1+1/\vartheta)} - 1)\right), \; j \in \{2, \ldots, d\}.$$

This, in conjunction with Theorem 1.8.1, allows us to implement the conditional distribution method for sampling Clayton copulas.

Tables 3.2 and 3.3 show the results we obtained for generating 100 000 vectors of random variates, where we choose $\vartheta = 0.8$ as before, corresponding to a Kendall's tau of 0.2857. Table 3.2 indicates the correctness of the approaches. As mentioned above, the p-values are derived from 1 000 samples. For the Fixed Talbot algorithm, we use the cut-off point $x_{\max} = 10^8$ with corresponding function value $\tilde{F}(x) = 0.999996$. Concerning the run times for this Clayton copula, which are listed in Table 3.3 for different dimensions, the direct approach is the fastest in all dimensions. The conditional distribution method is also quite fast in low dimensions. However, this is an exceptional case for this approach, since we are sampling a Clayton copula, for which the generator derivates, as well as the conditional distributions and their inverses, are easy to access without applying numerical root-search algorithms. Moreover, we are dealing with an Archimedean copula, thus it is comparably easy to compute and invert the desired Quantities (3.15) without the need of applying numerical root-search algorithms. Note that even in such a perfect scenario, the conditional distribution method is outperformed by the Fixed Talbot algorithm with $M = 6$ in large dimensions.

3.3.4 An outer power Clayton copula

In this example, we sample an outer power Archimedean copula with generator $\psi(t^{1/\beta})$ in three, ten, and one hundred dimensions, where $\psi(t) = (1+t)^{-2}$, i.e., ψ generates the Clayton copula with parameter equal to 0.5. Note that this parametric family of Archimedean generators is completely monotone for any $\beta \in [1, \infty)$, see Proposition 2.1.5 (3). As an example, we

3 Numerically sampling nested Archimedean copulas

Algorithm	p-value	$\hat{\tau}_{12}$	$\hat{\tau}_{13}$	$\hat{\tau}_{23}$
Direct approach sampling $V \sim \Gamma(5/4, 1)$	0.5332	0.2856	0.2832	0.2805
Conditional distribution method	0.2257	0.2821	0.2861	0.2886
Fixed Talbot with $M = 6$	0.3017	0.2858	0.2835	0.2877

Table 3.2 Precision of the direct approach, the conditional distribution method, and the Fixed Talbot algorithm for generating 100 000 vectors of random variates from a trivariate Clayton copula.

d	Algorithm	κ
3	Direct approach sampling $V \sim \Gamma(5/4, 1)$	0.1161
3	Conditional distribution method	0.1569
3	Fixed Talbot with $M = 6$	5.0915
10	Direct approach sampling $V \sim \Gamma(5/4, 1)$	0.3745
10	Conditional distribution method	0.7127
10	Fixed Talbot with $M = 6$	5.3543
100	Direct approach sampling $V \sim \Gamma(5/4, 1)$	3.3143
100	Conditional distribution method	7.1857
100	Fixed Talbot with $M = 6$	8.1392
200	Direct approach sampling $V \sim \Gamma(5/4, 1)$	6.0531
200	Conditional distribution method	14.3182
200	Fixed Talbot with $M = 6$	11.0464

Table 3.3 Run times in seconds for the direct approach, the conditional distribution method, and the Fixed Talbot algorithm for generating 100 000 vectors of random variates from a Clayton copula.

3.3 Experiments and examples

choose $\beta = 1.5$. For accessing the corresponding Laplace-Stieltjes inverse F, we use the Fixed Talbot algorithm with $M = 6$. For the truncation point, we choose $x_{\max} = 10^{16}$, resulting in $\tilde{F}(x_{\max}) = 0.999996$. The left-hand side of Figure 3.1 shows a plot of $\tilde{F}$. We obtain 0.5052

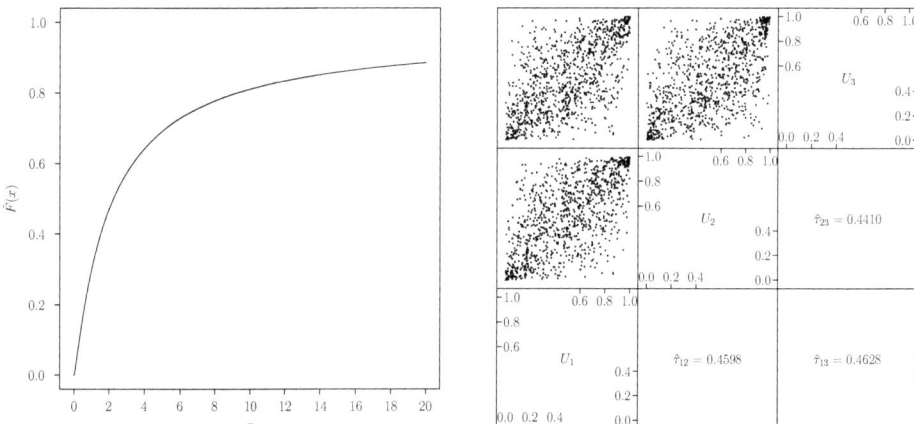

Figure 3.1 Approximation $\tilde{F}$ of F (left). A scatter plot matrix for the outer power Clayton copula with generator $(1 + t^{2/3})^{-2}$ (right).

as p-value of the Kolmogorov-Smirnov test for 1 000 generated samples of 100 000 vectors of random variates following this trivariate copula. Pairwise sample versions of Kendall's tau based on 100 000 generated vectors of random variates are given by $\hat{\tau}_{12} = 0.4667$, $\hat{\tau}_{13} = 0.4650$, and $\hat{\tau}_{23} = 0.4685$, which are close to the theoretical value 0.4666. Run times for generating 100 000 vectors for the three-, ten-, and one-hundred-dimensional case are given by 17.32s, 17.51s, and 22.54s, respectively. We note that the run time is mainly determined by the evaluation of $\tilde{F}$. The right-hand side of Figure 3.1 shows 1 000 generated vectors of random variates following this copula with corresponding sample versions of Kendall's tau.

3.3.5 A nested outer power Clayton copula

Assume a generator $\psi \in \Psi_\infty$ and $\beta_i \in [1, \infty)$, $i \in \{0, 1\}$, as given. It is easily verified that the two outer power generators $\psi(t^{1/\beta_0})$ and $\psi(t^{1/\beta_1})$ satisfy the nesting condition if $\beta_0 \leq \beta_1$. We call nested Archimedean copulas constructed in this way *nested outer power Archimedean copulas*.

3 Numerically sampling nested Archimedean copulas

In this example, we consider ψ from Clayton's family with parameter $\vartheta = 1$, i.e., $\psi(t) = 1/(1+t)$, and construct a nested outer power Archimedean copula with the two generators $\psi_i(t) = \psi(t^{1/\beta_i})$, $i \in \{0, 1\}$, with $\beta_0 = 1.1$ and $\beta_1 = 1.5$. As dimensions of the resulting nested outer power Clayton copula, we consider the cases 1(2), 5(5), and 50(50), where the notation "1(2)" refers to the trivariate fully-nested Archimedean copula of Type (2.5) and the cases "5(5)" and "50(50)" are constructed accordingly, simply involving larger dimensions for the non-sectorial and the sector part. For the resulting copulas, we apply the Fixed Talbot algorithm with $M = 6$ and $x_{\max} = 10^8$, corresponding to $\tilde{F}_0(x_{\max}) = 0.999996$, for sampling F_0. The left-hand side of Figure 3.2 shows a plot of $\tilde{F}_0$. As the inner distribution function F_{01} corresponds to a known stable distribution, we do not need to apply numerical inversion algorithms for sampling F_{01}. As p-value of the Kolmogorov-Smirnov test for the trivariate copula we obtain

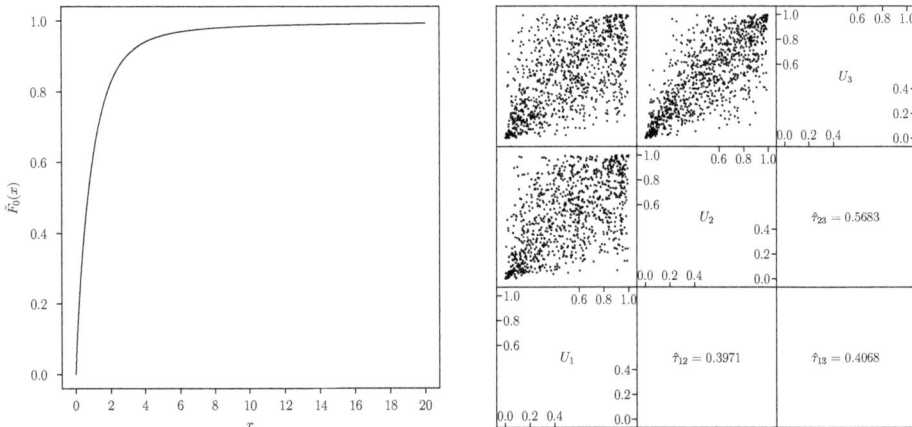

Figure 3.2 Approximation $\tilde{F}_0$ of F_0 (left). A scatter plot matrix for the nested outer power Clayton copula with generators $1/(1 + t^{1/\beta_i})$ and $\beta_i \in \{1.1, 1.5\}$ (right).

0.8103 based on 1 000 generated samples of 100 000 vectors of random variates. Pairwise sample versions of Kendall's tau based on 100 000 generated vectors of random variates are given by $\hat{\tau}_{12} = 0.3971$, $\hat{\tau}_{13} = 0.4068$, and $\hat{\tau}_{23} = 0.5683$, where the corresponding population versions are $\tau_{12} = \tau_{13} = 0.3939$ and $\tau_{23} = 0.5556$. Run times for generating 100 000 observations of the three-, ten-, and one-hundred-dimensional nested outer power Clayton copulas are given by 10.71s, 11.23s, and 17.52s, respectively. The right-hand side of Figure 3.2 shows 1 000 generated vectors of random variates following this copula with corresponding sample versions of Kendall's tau.

3.3.6 Nested Gumbel and Clayton copulas

For sampling a nested Gumbel copula, the inner generator is given by $\psi_{01}(t; V_0) = \exp(-V_0 t^\alpha)$, $\alpha = \vartheta_0/\vartheta_1$, which corresponds to a $S(\alpha, 1, (\cos(\alpha\pi/2)V_0)^{1/\alpha}, \mathbb{1}_{\{\alpha=1\}}; 1)$ distribution. This generator ψ_{01} is of the form $\psi(V_0^{1/\alpha})$ where ψ corresponds to a $S(\alpha, 1, \cos^{1/\alpha}(\alpha\pi/2), \mathbb{1}_{\{\alpha=1\}}; 1)$ distribution. Since $\psi(ct)$ generates the same copula as ψ for all $c \in (0, \infty)$, one may simply sample a Gumbel copula with generator ψ in Step (2) of McNeil's algorithm. Sampling a nested Gumbel copula is therefore particularly easy. First, only stable distributions are involved, for which there are direct sampling algorithms available. Second and most importantly, the stable distributions involved have fixed parameters, i.e., they are not affected by random variates obtained from earlier steps of the sampling algorithm.

Although there is no need to apply algorithms for numerically inverting Laplace transforms in order to sample nested Gumbel copulas, one could investigate such algorithms in this case, since only distributions with fixed parameters are involved. This boils down to investigating the precision with which a $S(1/\vartheta, 1, \cos^\vartheta(\pi/(2\vartheta)), \mathbb{1}_{\{\vartheta=1\}}; 1)$ distribution function F with different values $\vartheta \in [1, \infty)$ can be accessed. For this, we consider the parameters $\vartheta \in \{1.2, 1.6, 2.0, 2.4, 2.8, 3.2\}$ and evaluate $\tilde{F}(x)$ on an equidistant grid of $10\,000$ points x ranging from 0.001 to 2. For $\vartheta = 1.2$, the Fixed Talbot algorithm with $M = 6$ leads to unreliable results for all x considered and shows an oscillating behavior for x-values less than 0.47. The same behavior can be observed for $\vartheta = 1.6$ and x-values smaller than 0.11. For $\vartheta \geq 2$, the oscillation of $\tilde{F}(x)$ around zero in the neighborhood of 0 is negligible and one obtains reliable approximations of F. However, $\vartheta \geq 2$ implies a Kendall's tau greater than or equal to 0.5, which might be unrealistic in applications. Neither other choices of M nor considering the inversion problem $\mathcal{L}[\exp(-x)F(x)](t) = \mathcal{L}[F](t+1) = \psi(t+1)/(t+1)$ lead to smaller approximation errors. Note that the Gaver-Wynn-rho algorithm does not even show reliable results for any of the investigated parameters ϑ or x-values. The problem for numerical inversion algorithms seems to be the rather flat structure of F in the neighborhood of 0 for small ϑ, see the left-hand side of Figure 3.3, obtained with the statistical software package R.

The problem becomes even more severe when sampling nested Clayton copulas. In this case, $\psi_{01}(t; V_0) = \exp(-V_0((1+t)^\alpha - 1))$, $\alpha = \vartheta_0/\vartheta_1$, and thus the corresponding distribution function F_{01} is affected by the additional variable V_0. Note that V_0, obtained from Step (1) of McNeil's algorithm, changes almost surely for every vector of random variates to be drawn. This makes it difficult to obtain reliable approximations $\tilde{F}_{01}$ of F_{01}. For example, evaluating F_{01} for $\alpha \in \{0.2, 0.5, 0.8\}$ and $V_0 \in \{2, 5, 10\}$ using the Fixed Talbot algorithm with $M = 6$ leads to the following results. For $\alpha = 0.2$, the algorithm works fine for all $V_0 \in \{2, 5, 10\}$. For

3 Numerically sampling nested Archimedean copulas

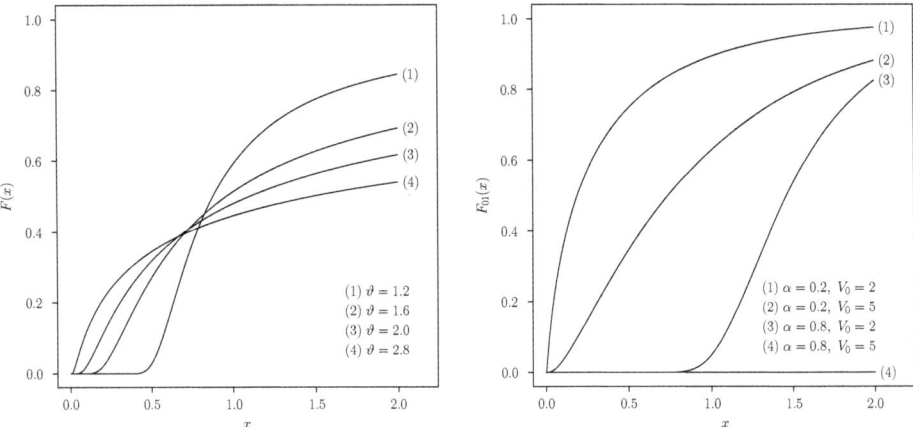

Figure 3.3 F corresponding to $S(1/\vartheta, 1, \cos^\vartheta(\pi/(2\vartheta)), \mathbb{1}_{\{\vartheta=1\}}; 1)$ (left). F_{01} corresponding to $\psi_{01}(t; V_0) = \exp(-V_0((1+t)^\alpha - 1))$ (right).

$\alpha = 0.5$, the approximation $\tilde{F}_{01}$ shows negative values, however, negligibly small in absolute value. These negative values appear more often if V_0 increases from 2 to 10. For $\alpha = 0.8$, the oscillating behavior of $\tilde{F}_{01}$ appears again and does not lead to a reliable approximation of F_{01} for any $V_0 \in \{2, 5, 10\}$. Note that a small α for a nested Clayton copula translates to a rather weak dependence between components belonging to different sectors and rather strong dependence between components in the same sector. This may be not realistic for the model in mind. Even if α is small, there is no guarantee that $\tilde{F}_{01}$ leads to reliable results when V_0 is large. One can infer from the right-hand side of Figure 3.3, again obtained with the statistical software package R, that the distribution function F_{01} corresponding to ψ_{01} gets more and more flat the larger α and V_0 are. Due to the almost surely changing parameter V_0 affecting the structure of F_{01}, sampling F_{01} for a nested Clayton copula with the Fixed Talbot algorithm is therefore problematic.

3.3.7 An Ali-Mikhail-Haq(Clayton) copula

In this example, we build and sample nested Archimedean copulas of the form "A(C)", i.e., a Clayton copula is nested into an Ali-Mikhail-Haq copula, in the dimensions 1(2), 5(5), and 50(50) as before. For the Ali-Mikhail-Haq generator ψ_0, we choose $\vartheta_0 = 0.8$, and for the Clayton generator ψ_1, we use $\vartheta_1 = 2$. For the resulting nested Archimedean copula, we directly sample the known distribution function F_0 in Step (1) of McNeil's algorithm and use the Fixed Talbot algorithm with $M = 6$ and $x_{\max} = 10^4$, corresponding to $\tilde{F}_{01}(x_{\max}) = 0.999996$. The left-hand side of Figure 3.4 shows plots of $\tilde{F}_{01}$ for the samples $V_0 \in \{1, 2, 5, 10\}$ of F_0. Based on the same sample sizes as before, the p-value of the Kolmogorov-Smirnov test for the trivariate copula is reported as 0.5322. Sample versions of Kendall's tau are given by $\hat{\tau}_{12} = 0.2349$, $\hat{\tau}_{13} = 0.2358$, and $\hat{\tau}_{23} = 0.5029$, corresponding to the population versions $\tau_{12} = \tau_{13} = 0.2337$ and $\tau_{23} = 0.5$. Run times for generating 100 000 vectors of random variates of the three-, ten-, and one-hundred-dimensional nested A(C) copulas are given by 7.91s, 8.19s, and 11.72s, respectively. The right-hand side of Figure 3.4 shows a scatter plot matrix with corresponding sample versions of Kendall's tau of the trivariate nested A(C) copula. We particularly emphasize the different tail behavior of the generated data, see also Table 2.2.

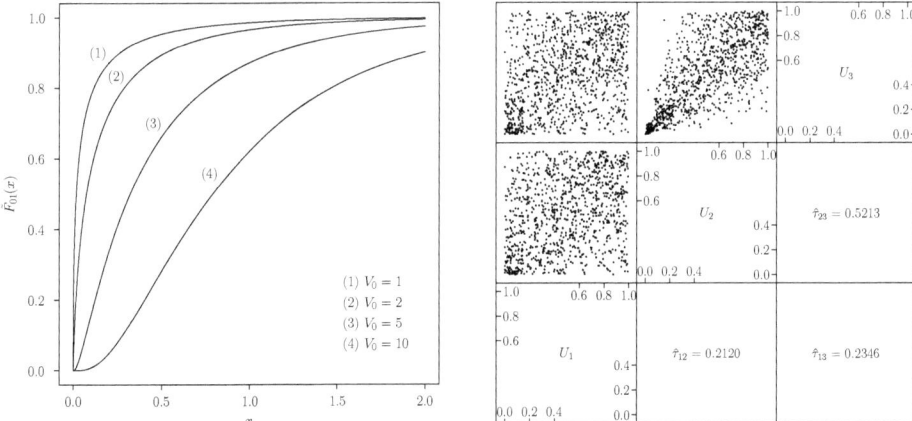

Figure 3.4 Approximations $\tilde{F}_{01}$ of F_{01} (left). A scatter plot matrix for the A(C) copula with generators $0.2/(\exp(t) - 0.8)$ and $(1 + t)^{-1/2}$ (right).

3.4 Conclusion

Generating random variates from Archimedean copulas with the standard conditional distribution method gets more and more complicated as the dimension increases, due to the high-order generator derivatives. These derivatives are usually not easy to access. Moreover, due to

3 Numerically sampling nested Archimedean copulas

the mixed partial derivatives involved, it is practically impossible to apply the conditional distribution method for sampling nested Archimedean copulas. Both Archimedean and nested Archimedean copulas can be sampled efficiently with the algorithms of Marshall and Olkin (1988) and McNeil (2008). Sampling an Archimedean copula with generator ψ with the former algorithm requires sampling the distribution function $F = \mathcal{LS}^{-1}[\psi]$. Sampling a nested Archimedean copula based on the generators ψ_i, $i \in \{0, 1\}$, with the latter algorithm involves sampling the distribution functions $F_0 = \mathcal{LS}^{-1}[\psi_0]$ and $F_{01} = \mathcal{LS}^{-1}[\psi_{01}(\cdot\,;V_0)]$, where $V_0 \sim F_0$.

Drawing random variates from F_0 and F_{01} when there are no explicit strategies known can be achieved with numerical inversion of Laplace transforms. This is motivated by the fact that only one random variate from F, respectively each of F_0 and F_{01}, has to be drawn for generating one vector of random variates following the corresponding copula, independently of the number of dimensions. As algorithms for numerically inverting Laplace transforms, we investigated the Fixed Talbot, Gaver-Stehfest, Gaver-Wynn-rho, and Laguerre-series algorithm regarding precision and run time. In particular, the Fixed Talbot algorithm turned out to be fast and highly accurate for several examples considered. All in all, procedures for numerical inversion of Laplace transforms may be efficiently used for sampling Archimedean copulas. Since F is a distribution function with fixed parameters, it is comparably easy to carry out quality checks beforehand. This is also important insofar as verifying the assumptions on which a numerical algorithm is based is usually difficult, e.g., one usually does not know beforehand if F is indeed continuous.

Sampling a nested Archimedean copula is much more involved. The main problem is that the distribution function F_{01} not only depends on the parameters of ψ_0 and ψ_1, but also on the random variate $V_0 \sim F_0$ obtained from the first step of the sampling algorithm. At first glance, V_0 can be seen as an additional parameter affecting F_{01}. However, if F_0 is absolutely continuous, then V_0 almost surely changes with every vector of random variates to be generated. Indeed, except for certain cases, e.g., a nested Gumbel copula, sampling n vectors of random variates following a nested Archimedean copula with McNeil's algorithm does not only require sampling n random variates from the distribution F_0 with fixed parameters, but also sampling n random variates from the univariate distribution F_{01} with, possibly almost surely, different parameters for each of the variates to be drawn. For this reason, efficiently sampling F_{01} is usually considerably more complicated than sampling F_0. In order to ensure that a numerical procedure works accurately for any random variate V_0, careful checks of the generated data should be made. This is usually hard, if not impossible, to accomplish and time-consuming at best.

3.4 Conclusion

With this in mind, we close this section by advising the reader to use numerical inversion algorithms for sampling purposes with care. Such methods are usually applied when we do not know the underlying inverse Laplace-Stieltjes transforms. We most likely do not even know if they are continuous and numerical inversion procedures often have difficulties if they are not. Moreover, there is no guarantee that algorithms which worked well in the our examples are applicable to others and so an efficient sampling algorithm is always linked to the specific Archimedean generator(s) underlying.

3 Numerically sampling nested Archimedean copulas

4 Directly sampling nested Archimedean copulas

"Eureka!"
Archimedes (-0286 to -0211)

In this chapter, we present efficient sampling algorithms for nested Archimedean copulas. In contrast to the approach taken in the last chapter based on numerical inversion of Laplace transforms and to avoid possible difficulties arising from such procedures, our goal is to develop direct sampling strategies for the inverse Laplace-Stieltjes transforms F_0 and F_{01} involved in a nested Archimedean copula. Note that the class of inner distribution functions F_{01} precisely coincides with the class of infinitely divisible distributions, see Feller (1971, p. 450). Approximate sampling methods for such distributions, when the Lévy measure associated with the Laplace-Stieltjes transform is known, have been presented in Bondesson (1982) and Damien et al. (1995). In the approaches we present, the main key to fast sampling algorithms is to significantly reduce the influence of the random variable $V_0 \sim F_0$ on the inner distribution function F_{01}. Further, our sampling strategies lead to exact algorithms.

As a first result, we provide efficient sampling algorithms for the distributions involved in sampling the nested Archimedean families of Ali-Mikhail-Haq, Frank, and Joe. The idea behind the presented algorithms readily applies to other distributions showing similar characteristics. The main key here is that F_0 and F_{01} are discrete with jumps at $\mathbb{N}$. Second, we introduce a general strategy on how to build nested Archimedean copulas from given Archimedean generators. Sampling these copulas involves sampling exponentially tilted stable distributions. For this task, we develop a fast rejection algorithm. It is proven for the more general class of tilted Archimedean generators that this algorithm reduces the complexity of the standard rejection algorithm to logarithmic complexity. These results are crucial, e.g., for sampling nested Clayton copulas. Third, with the additional help of randomization of generator parameters, we find explicit sampling algorithms for several nested Archimedean copulas based on different Archimedean

4 Directly sampling nested Archimedean copulas

families. Finally, we give detailed examples including remarks about the implementation.

The results and sampling strategies presented allow us to construct and sample nested Archimedean copulas. As all algorithms work fast for large parameter ranges and do not require numerical inversion of Laplace transforms, we can recommend them for large-scale simulation studies, even in large dimensions, see Chapter 5. The results presented in this chapter have been submitted for publication.

4.1 Nested Ali-Mikhail-Haq, Frank, and Joe copulas

For an Archimedean generator $\psi \in \Psi_\infty$, $F = \mathcal{LS}^{-1}[\psi]$ may be a discrete distribution function. In such a case, numerical inversion of Laplace-Stieltjes transforms may be problematic. Fortunately, the case of a discrete F is precisely covered by the following result from which one can often deduce the jump locations and jump heights of a discrete F for a given ψ.

Theorem 4.1.1
Let $\psi \in \Psi_\infty$ with $F = \mathcal{LS}^{-1}[\psi]$ and let $G(x) := \sum_{k=1}^{\infty} p_k \mathbb{1}_{[x_k, \infty)}(x)$, with $0 < x_1 < x_2 < \ldots$ and $p_k \geq 0$, $k \in \mathbb{N}$, with $\sum_{k=1}^{\infty} p_k = 1$. Then

$$F = G \Leftrightarrow \psi(t) = \sum_{k=1}^{\infty} p_k \exp(-x_k t), \ t \in [0, \infty].$$

Proof
The "only if" part of the claim follows from the definition of Laplace-Stieltjes transforms. For the "if" part of the statement, note that the Laplace-Stieltjes transform of G at $t \in [0, \infty]$ equals $\sum_{k=1}^{\infty} p_k \exp(-x_k t)$, which in turn equals ψ, the Laplace-Stieltjes transform of F. The Uniqueness Theorem implies that $F = G$ up to a set of Lebesgue measure zero, which implies $F = G$, since F and G are right-continuous. □

An application of Theorem 4.1.1 leads to the probability mass functions of F for the families of Ali-Mikhail-Haq, Frank, and Joe as listed in Table 2.1. The computation involves a geometric series expansion for Ali-Mikhail-Haq's, a logarithmic series expansion for Frank's, and a binomial series expansion for Joe's generator. Note that these families are lower tail independent, see Proposition 2.3.3 (2).

If the outer distribution function $F = \mathcal{LS}^{-1}[\psi]$ is discrete, one may ask if there are also cases in which the inner distribution function $F_{01} = \mathcal{LS}^{-1}[\psi_{01}]$ is discrete. The following result provides the inner distribution functions F_{01} for the nested Archimedean families of Ali-Mikhail-Haq, Frank, and Joe and positively answers this question.

4.1 Nested Ali-Mikhail-Haq, Frank, and Joe copulas

Theorem 4.1.2

(1) If $\psi_i(t) := (1 - \vartheta_i)/(\exp(t) - \vartheta_i)$, $t \in [0, \infty]$, with $\vartheta_i \in [0, 1)$, $i \in \{0, 1\}$, such that $\vartheta_0 \leq \vartheta_1$, then $\psi_{01}(t; V_0)$, with $V_0 \in \mathbb{N}$, has inverse Laplace-Stieltjes transform $F_{01}(x) = \sum_{k=V_0}^{\infty} p_k \mathbb{1}_{[x_k, \infty)}(x)$ with

$$x_k := k, \quad p_k := \frac{c_1^{k-V_0}}{c_0^k}\binom{k-1}{k-V_0}, \quad k \in \{V_0, V_0 + 1, \dots\},$$

where $c_0 := (1 - \vartheta_0)/(1 - \vartheta_1)$ and $c_1 := (\vartheta_1 - \vartheta_0)/(1 - \vartheta_1)$. Thus, for the Archimedean family of Ali-Mikhail-Haq, F_{01} is discrete with mass function p_k at $k \in \{V_0, V_0 + 1, \dots\}$.

(2) If $\psi_i(t) := -\log(1 - (1 - e^{-\vartheta})\exp(-t))/\vartheta$, $t \in [0, \infty]$, with $\vartheta_i \in (0, \infty)$, $i \in \{0, 1\}$, such that $\vartheta_0 \leq \vartheta_1$, then $\psi_{01}(t; V_0)$, with $V_0 \in \mathbb{N}$, has inverse Laplace-Stieltjes transform $F_{01}(x) = \sum_{k=V_0}^{\infty} p_k \mathbb{1}_{[x_k, \infty)}(x)$ with

$$x_k := k, \quad p_k := \frac{c_1^k}{c_0^{V_0}} \sum_{j=0}^{V_0} \binom{V_0}{j}\binom{j\vartheta_0/\vartheta_1}{k}(-1)^{j+k}, \quad k \in \{V_0, V_0 + 1, \dots\},$$

where $c_i = 1 - e^{-\vartheta_i}$ for $i \in \{0, 1\}$. Thus, for the Archimedean family of Frank, F_{01} is discrete with probability density function p_k at $k \in \{V_0, V_0 + 1, \dots\}$.

(3) If $\psi_i(t) := 1 - (1 - \exp(-t))^{1/\vartheta_i}$, $t \in [0, \infty]$, with $\vartheta_i \in [1, \infty)$, $i \in \{0, 1\}$, such that $\vartheta_0 \leq \vartheta_1$, then $\psi_{01}(t; V_0)$, with $V_0 \in \mathbb{N}$, has inverse Laplace-Stieltjes transform $F_{01}(x) = \sum_{k=V_0}^{\infty} p_k \mathbb{1}_{[x_k, \infty)}(x)$ with

$$x_k := k, \quad p_k := \sum_{j=0}^{V_0} \binom{V_0}{j}\binom{j\vartheta_0/\vartheta_1}{k}(-1)^{j+k}, \quad k \in \{V_0, V_0 + 1, \dots\}.$$

Thus, for the Archimedean family of Joe, F_{01} is discrete with probability density function p_k at $k \in \{V_0, V_0 + 1, \dots\}$.

Proof

For (1), an application of the Binomial Theorem leads to

$$\begin{aligned}
\psi_{01}(t; V_0) &= (c_0 \exp(t))^{-V_0}(1 - c_1 \exp(-t)/c_0)^{-V_0} \\
&= (c_0 \exp(t))^{-V_0} \sum_{k=0}^{\infty} \binom{-V_0}{k}(-1)^k (c_1 \exp(-t)/c_0)^k \\
&= \sum_{k=0}^{\infty} c_0^{-(V_0+k)} c_1^k \binom{V_0 + k - 1}{k} \exp(-(V_0 + k)t) \\
&= \sum_{k=V_0}^{\infty} c_0^{-k} c_1^{k-V_0} \binom{k-1}{k-V_0} \exp(-kt), \quad t \in [0, \infty].
\end{aligned}$$

4 Directly sampling nested Archimedean copulas

Trivially, $\sum_{k=V_0}^{\infty} p_k = 1$ as $\psi_{01}(0; V_0) = 1$. Thus the stated result follows from Theorem 4.1.1. For (2), let $\alpha := \vartheta_0/\vartheta_1$ and apply the Binomial Theorem twice to see that

$$\psi_{01}(t; V_0) = c_0^{-V_0}\left(1 - (1 - c_1 \exp(-t))^{\alpha}\right)^{V_0}$$

$$= c_0^{-V_0} \sum_{j=0}^{V_0} \binom{V_0}{j}\left(-(1 - c_1 \exp(-t))^{\alpha}\right)^{j}$$

$$= \sum_{j=0}^{V_0} c_0^{-V_0} \binom{V_0}{j}(-1)^j \sum_{k=0}^{\infty} \binom{j\alpha}{k}(-c_1 \exp(-t))^k$$

$$= \sum_{k=0}^{\infty} \left(\frac{c_1^k}{c_0^{V_0}} \sum_{j=0}^{V_0} \binom{V_0}{j}\binom{j\alpha}{k}(-1)^{j+k}\right) \exp(-kt), \ t \in [0, \infty].$$

Let p_k be as in the claim, for all $k \in \mathbb{N}_0$. It remains to show that $p_k \geq 0$, $k \in \mathbb{N}_0$, and that $p_k = 0$ for $k \in \{0, \ldots, V_0 - 1\}$. It suffices to show these statements for $\tilde{p}_k := \sum_{j=0}^{V_0} \binom{V_0}{j}\binom{j\alpha}{k}(-1)^{j+k}$, $k \in \mathbb{N}_0$. Again by the Binomial Theorem, the generating function of $(\tilde{p}_k)_{k \in \mathbb{N}_0}$ can be calculated as $g(x) := (1 - (1 - x)^{\alpha})^{V_0}$. Now g is absolutely monotone as a composition of the two absolutely monotone functions x^{V_0} and $1 - (1 - x)^{\alpha}$. Thus, $\tilde{p}_k \geq 0$, $k \in \mathbb{N}_0$. Further, $\tilde{p}_k = g^{(k)}(x)/k!|_{x=0}$. For showing that $\tilde{p}_k = 0$ for $k \in \{0, \ldots, V_0 - 1\}$, it therefore suffices to show that $g^{(k)}(x)|_{x=0} = 0$ for $k \in \{0, \ldots, V_0 - 1\}$ and all $V_0 \in \mathbb{N}$. We proceed by induction. The case $V_0 = 1$ is trivial. Assume the statement holds for $V_0 \in \mathbb{N}$ and consider the case $V_0 + 1$. Applying the Leibniz Rule leads to

$$\frac{d^k}{dx^k}(1 - (1 - x)^{\alpha})^{V_0 + 1} = \frac{d^k}{dx^k}\left((1 - (1 - x)^{\alpha})^{V_0}(1 - (1 - x)^{\alpha})\right)$$

$$= \sum_{j=0}^{k} \binom{k}{j}\left((1 - (1 - x)^{\alpha})^{V_0}\right)^{(j)}(1 - (1 - x)^{\alpha})^{(k-j)}.$$

By the induction hypothesis, $\frac{d^j}{dx^j}(1 - (1 - x)^{\alpha})^{V_0}|_{x=0} = 0$ for all $j \in \{0, \ldots, V_0 - 1\}$, thus, $\frac{d^k}{dx^k}(1 - (1 - x)^{\alpha})^{V_0 + 1}|_{x=0} = 0$ for all $k \in \{0, \ldots, V_0 - 1\}$ and the claim follows. Part (3) works similarly as Part (2) and is therefore omitted. □

Another question is to ask how discrete inner distribution functions F_{01} can be sampled efficiently. As a first guess, one might be tempted to precalculate and store the function values of F_{01} for the different variates V_0 obtained from Step (1) of McNeil's algorithm. However, the efficiency of this procedure strongly depends on the number of different values V_0 one expects to obtain. For example, if $\mathbb{E}[V_0] = \infty$, as, e.g., for the Archimedean family of Joe, one has to expect different and rather large random variates V_0 in a computational implementation. Further, it may be numerically challenging to evaluate the jump heights of F_{01} as they depend

4.1 Nested Ali-Mikhail-Haq, Frank, and Joe copulas

on V_0. As one can infer from Theorem 4.1.2 (3), the jump heights of F_{01} for a nested Joe copula are given by sums of numerically challenging products of binomial coefficients. The first jump of F_{01} occurs at V_0 with height $p_{V_0} = \alpha^{V_0}$, $\alpha := \vartheta_0/\vartheta_1$, and numerical experiments indicate that computing the jump heights becomes especially demanding if V_0 is large or α is small.

Although one usually tries to figure out how F_{01} looks like and then aims for an efficient sampling algorithm, a more sophisticated approach works with the Laplace-Stieltjes transform directly. The idea is to write ψ_{01} as a power in V_0 via

$$\psi_{01}(\,\cdot\,;V_0) = \exp(-V_0\psi_0^{-1} \circ \psi_1) = (\exp(-\psi_0^{-1} \circ \psi_1))^{V_0}. \tag{4.1}$$

Whenever $V_0 \in \mathbb{N}$, one may infer from Proposition A.3.11 that F_{01} is the distribution function of a V_0-fold sum of independent and identically distributed random variables following the distribution $F := \mathcal{LS}^{-1}[\exp(-\psi_0^{-1} \circ \psi_1)]$. The main advantage is that the random variable V_0 of F_0 now enters F_{01} only in the number of summands to be sampled, but does not affect F. One can therefore decide about an efficient sampling algorithm according to the specific form and properties of F. One could also bring numerical inversion of Laplace transforms into play, as F can be investigated for fixed parameters beforehand and careful quality checks can be made.

The following result presents direct sampling strategies for F_{01} for nested Ali-Mikhail-Haq, Frank, and Joe copulas based on the underlying distribution function F.

Theorem 4.1.3

(1) For the family of Ali-Mikhail-Haq, F_{01} can be sampled via the following algorithm, where V_0 denotes a random variate drawn from F_0.

  (1.1) Sample i.i.d. $V_{01,k} \sim \text{Geo}((1-\vartheta_1)/(1-\vartheta_0))$, $k \in \{1,\ldots,V_0\}$.

  (1.2) Return $V_{01} = \sum_{k=1}^{V_0} V_{01,k}$.

(2) For the family of Frank, F_{01} can be sampled via the following algorithm, where V_0 denotes a random variate drawn from F_0.

  (2.1) Sample i.i.d. $V_{01,k}$, $k \in \{1,\ldots,V_0\}$, with mass function given by $p_k = \binom{\vartheta_0/\vartheta_1}{k}(-1)^{k-1}(1-e^{-\vartheta_1})^k/(1-e^{-\vartheta_0})$ at $k \in \mathbb{N}$.

  (2.2) Return $V_{01} = \sum_{k=1}^{V_0} V_{01,k}$.

(3) For the family of Joe, F_{01} can be sampled via the following algorithm, where V_0 denotes a random variate drawn from F_0.

4 Directly sampling nested Archimedean copulas

(3.1) Sample i.i.d. $V_{01,k}$, $k \in \{1, \ldots, V_0\}$, with mass function given by $p_k = \binom{\vartheta_0/\vartheta_1}{k}(-1)^{k-1}$ at $k \in \mathbb{N}$.

(3.2) Return $V_{01} = \sum_{k=1}^{V_0} V_{01,k}$.

Proof

By Proposition A.3.11 and (4.1), it suffices to show that the random variables $V_{01,k}$, $k \in \{1, \ldots, V_0\}$, follow the correct distributions $F = \mathcal{LS}^{-1}[\exp(-\psi_0^{-1} \circ \psi_1)]$. For (1), set $p := (1 - \vartheta_1)/(1 - \vartheta_0)$ and verify that

$$\exp\bigl(-\psi_0^{-1}(\psi_1(t))\bigr) = \frac{p}{\exp(t) - (1-p)} = \frac{p \exp(-t)}{1 - (1-p)\exp(-t)}$$

$$= p \exp(-t) \sum_{k=0}^{\infty}(1-p)^k \exp(-kt) = \sum_{k=1}^{\infty} p(1-p)^{k-1} \exp(-kt)$$

for all $t \in [0, \infty]$. Thus, the claim follows from Theorem 4.1.1. For (2), the Binomial Theorem implies that

$$\exp\bigl(-\psi_0^{-1}(\psi_1(t))\bigr) = \frac{1}{1 - e^{-\vartheta_0}}\Bigl(1 - (1 - (1 - e^{-\vartheta_1})\exp(-t))^{\vartheta_0/\vartheta_1}\Bigr)$$

$$= \frac{1}{1 - e^{-\vartheta_0}}\Bigl(1 - \sum_{k=0}^{\infty}\binom{\vartheta_0/\vartheta_1}{k}(-1)^k(1 - e^{-\vartheta_1})^k \exp(-kt)\Bigr)$$

$$= \sum_{k=1}^{\infty}\binom{\vartheta_0/\vartheta_1}{k}(-1)^{k-1}(1 - e^{-\vartheta_1})^k \exp(-kt)/(1 - e^{-\vartheta_0})$$

for all $t \in [0, \infty]$. Thus, the claim follows from Theorem 4.1.1. Part (3) works similarly as Part (2) and is therefore omitted. □

Sampling F_{01} for a nested Ali-Mikhail-Haq copula can be achieved in a fast way, since sampling $F = \mathcal{LS}^{-1}\bigl[\exp(-\psi_0^{-1} \circ \psi_1)\bigr]$ requires only drawing geometric random variates, as for F_0. For sampling F for Frank's family, no specific strategy is known. One option is therefore to precalculate and store the sums of the jump heights $(p_k)_{k \in \mathbb{N}}$ of F up to $1 - \varepsilon$ for some sufficiently small $\varepsilon > 0$. For most parameter choices the corresponding algorithms are very fast, even if F grows slowly. For Joe's family, F is again of the same type as F_0, i.e., a distribution function which puts mass $p_k = \binom{\alpha}{k}(-1)^{k-1}$ on $k \in \mathbb{N}$. There is also no specific sampling strategy known for this distribution. In contrast to Frank's family, however, we are able to compute F at $n \in \mathbb{N}$ without knowing the function values at $k \in \{1, \ldots, n-1\}$, see the following result, where $\mathrm{B}(x,y) := \int_0^1 t^{x-1}(1-t)^{y-1} dt = \Gamma(x)\Gamma(y)/\Gamma(x+y)$ denotes the *beta function* at $x, y \in (0, \infty)$.

Proposition 4.1.4

Let F be the discrete distribution function of the distribution which puts mass $p_k = \binom{\alpha}{k}(-1)^{k-1}$

at $k \in \mathbb{N}$, where $\alpha \in (0,1)$. Then,

$$F(n) = 1 - (-1)^n \binom{\alpha-1}{n} = 1 - \frac{1}{n\,\mathrm{B}(n, 1-\alpha)}, \quad n \in \mathbb{N}.$$

Proof

$F(n) = \sum_{k=1}^{n} p_k = 1 - \sum_{k=0}^{n} \binom{\alpha}{k}(-1)^k$. An induction shows that the latter sum equals $(-1)^n \binom{\alpha-1}{n}$. By iteratively applying the functional equation for the gamma function, one obtains $\prod_{j=1}^{n}(j-\alpha) = \Gamma(n-\alpha+1)/\Gamma(1-\alpha)$. Applying this to the binomial coefficient $\binom{\alpha-1}{n}$ and using $n! = n\Gamma(n)$ leads to $F(n) = 1 - \Gamma(n-\alpha+1)/(n\Gamma(n)\Gamma(1-\alpha))$, thus the claim holds. □

Proposition 4.1.4 can be applied to sample a small amount of random variates or for finding U-quantiles for U close to one. For large amounts of random variates or for sampling the core of F, a table-lookup method may be applied. Since table-lookup methods are fast, they are useful in this setup as one usually requires large amounts of summands to be drawn from F in order to obtain a random variate $V_{01} \sim F_{01}$. Further, setting up tables can be done for fixed ϑ_0 and ϑ_1 beforehand, independently of V_0. Note that sampling from such a table can be achieved without truncation errors, see Subsection 3.3.1.

4.2 Nested Archimedean copulas based on generator transformations

Apart from the parametric Archimedean families listed in Table 2.1, there are not many Archimedean families with explicitly given generator inverses known. In order to obtain new Archimedean generators, one could try to transform given generators. Regarding sampling algorithms, such transformations will be most convenient if one knows how the corresponding distribution changes. This section presents two of such transformations. The first transformation consists of a simple tilt of a given Archimedean generator. The tilted generator corresponds to an exponentially tilted distribution. We introduce a fast rejection algorithm for sampling such distributions. The second transformation starts with an Archimedean generator and transforms it in such a way that nested Archimedean copulas result. Further, we derive stochastic representations of the random variables of the outer and inner distribution functions F_0 and F_{01}, respectively. As they both involve exponentially tilted stable distributed random variables, the presented fast rejection algorithm can be applied for efficiently sampling the resulting nested Archimedean copulas.

4 Directly sampling nested Archimedean copulas

4.2.1 Exponentially tilted generators and a fast sampling algorithm

One of the most simple transformations can be obtained by shifting the graph of an Archimedean generator ψ to the left and appropriately scaling it afterwards so that it is one at zero and therefore in Ψ_∞. The resulting *tilted Archimedean generator* $\tilde{\psi}$ will play an important role in the sequel and is addressed in the following result.

Theorem 4.2.1

Let $\psi \in \Psi_\infty$ with $F := \mathcal{LS}^{-1}[\psi]$ and $\tilde{\psi}(t) := \psi(h+t)/\psi(h)$, $t \in [0, \infty]$, where $h \in [0, \infty)$. Then, $\tilde{\psi} \in \Psi_\infty$ and the inverse Laplace-Stieltjes transform $\tilde{F} := \mathcal{LS}^{-1}[\tilde{\psi}]$ of $\tilde{\psi}$ is given by

$$\tilde{F}(x) = \int_0^x \exp(-hy)\, dF(y)/\psi(h), \ x \in [0, \infty). \tag{4.2}$$

Proof

As ψ is completely monotone, so is $\tilde{\psi}$. Further, $\tilde{\psi}(0) = 1$. Thus, $\tilde{\psi} \in \Psi_\infty$. Let $G(x)$ denote the right-hand side of (4.2). The Integration and Differentiation Theorem, Corollary 2.1.9, and the Fubini-Tonelli Theorem, see Folland (1999, p. 67), imply

$$\mathcal{LS}[G](t) = t\mathcal{L}[G](t) = \int_0^\infty t \exp(-tx) \int_0^x \exp(-hy)\, dF(y)\, dx/\psi(h)$$
$$= \int_0^\infty \exp(-hy) \int_y^\infty t \exp(-tx)\, dx\, dF(y)/\psi(h)$$
$$= \int_0^\infty \exp(-(h+t)y)\, dF(y)/\psi(h) = \psi(h+t)/\psi(h), \ t \in (0, \infty). \qquad \square$$

Concerning the implied dependence properties of Archimedean copulas constructed with tilted Archimedean generators, a simple substitution shows that Kendall's tau $\tilde{\tau}$ of the Archimedean copula generated by $\tilde{\psi}$ is given by

$$\tilde{\tau} = 1 - \frac{4}{(\psi(h))^2} \int_h^\infty (t-h)(\psi'(t))^2\, dt.$$

Now assume that the lower tail-dependence coefficient λ_l of the Archimedean copula generated by ψ, as well as the one corresponding to the copula generated by $\tilde{\psi}$, denoted by $\tilde{\lambda}_l$, exist. By the first equality in Formula (2.11), it follows that $\tilde{\lambda}_l \geq \lambda_l$. The second equality in Formula (2.12) implies that $\tilde{\lambda}_u = 0$.

If $F = \mathcal{LS}^{-1}[\psi]$ admits a density f, then Theorem 4.2.1 states that $\tilde{F} = \mathcal{LS}^{-1}[\tilde{\psi}]$ also admits a density, given by

$$\tilde{f}(x) = \exp(-hx)f(x)/\psi(h), \ x \in [0, \infty).$$

4.2 Nested Archimedean copulas based on generator transformations

The next question is to ask how we can sample such an *exponentially tilted distribution* with tilt h. If f is easy to sample, a general algorithm for sampling the density $\tilde{f}$ is the standard *rejection algorithm*, see Devroye (1986, pp. 40), with $f(x)/\psi(h)$ as envelope.

Algorithm 4.2.2 (Standard rejection)
Repeatedly sample $\tilde{V} \sim f$ and $U \sim \mathrm{U}[0,1]$ until $U \leq e^{-h\tilde{V}}$, then return $\tilde{V}$.

The expected number of iterations in the standard rejection algorithm is $1/\psi(h)$, see Devroye (1986, p. 42), i.e., the algorithm has complexity $\mathcal{O}(1/\psi(h))$ assuming constant complexity for sampling f. As we will see in Subsection 4.4.3, this may slow down the algorithm considerably, making it inapplicable for certain Archimedean families and parameter ranges.

Our goal is therefore to reduce the complexity of the standard rejection algorithm. For this task, we use a similar trick as in (4.1) and cleverly write the Laplace-Stieltjes transform $\tilde{\psi}$ as a product which can then be sampled as a sum of independent and identically distributed random variables. By choosing this product in such a way that the complexity of the corresponding sampling algorithm is minimized, we obtain the following result. Note that $\lfloor \cdot \rfloor$ denotes the floor function, as usual.

Theorem 4.2.3
Let $\psi \in \Psi_\infty$ such that $\psi^{1/m} \in \Psi_\infty$ for all $m \in \mathbb{N}$ and let $\tilde{\psi}(t) = \psi(h+t)/\psi(h)$, $t \in [0,\infty]$, where $h \in [0,\infty)$.

(1) Let $\tilde{V} := \sum_{k=1}^m \tilde{V}_k$ with i.i.d. $\tilde{V}_k \sim \tilde{F}_m := \mathcal{LS}^{-1}[\tilde{\psi}^{1/m}]$, $k \in \{1,\ldots,m\}$, for some $m \in \mathbb{N}$. Then $\tilde{V} \sim \tilde{F} = \mathcal{LS}^{-1}[\tilde{\psi}]$.

(2) Let $c := 1/\psi(h)$. By choosing m in (1) as

$$m = \begin{cases} 1, & \log(c) \leq 1, \\ \lfloor \log(c) \rfloor, & \log(c) > 1,\ \lfloor \log(c) \rfloor c^{1/\lfloor \log(c) \rfloor} \leq \lceil \log(c) \rceil c^{1/\lceil \log(c) \rceil}, \\ \lceil \log(c) \rceil, & \log(c) > 1,\ \lfloor \log(c) \rfloor c^{1/\lfloor \log(c) \rfloor} > \lceil \log(c) \rceil c^{1/\lceil \log(c) \rceil}, \end{cases} \quad (4.3)$$

and by applying the standard rejection algorithm for sampling each $\tilde{V}_k$, $k \in \{1,\ldots,m\}$, the expected number of iterations for sampling $\tilde{V}$ is less than or equal to $1/\psi(h)$. Further, if $\log(c) > 1$, this strategy for sampling $\tilde{F}$ has complexity $\mathcal{O}\big(\log(1/\psi(h))\big)$.

Proof
Part (1) follows from Proposition A.3.11. For (2), note that sampling $\tilde{F}_m$ m-times via the standard rejection algorithm is expected to take $mc^{1/m}$ iterations. For determining the choice of $m \in \mathbb{N}$ that minimizes this number, consider the function $g(x) = xc^{1/x}$ on $(0,\infty)$. Since g has

4 Directly sampling nested Archimedean copulas

its global minimum at $\log(c)$ and since g is convex, the optimal $m \in \mathbb{N}$ is either $m = \lfloor \log(c) \rfloor$ or $m = \lceil \log(c) \rceil$, chosen such that the expected number of iterations is minimized, with $m = 1$ if $\log(c) \leq 1$. If $\log(c) \leq 1$, the resulting expected number of iterations is simply $c = 1/\psi(h)$, as for the standard rejection algorithm. If $\log(c) > 1$, i.e., $c \in (e, \infty)$, it is $\min\{\lfloor \log(c) \rfloor c^{1/\lfloor \log(c) \rfloor}, \lceil \log(c) \rceil c^{1/\lceil \log(c) \rceil}\}$ and therefore less than or equal to $\lfloor \log(c) \rfloor c^{1/\lfloor \log(c) \rfloor}$. If $c \in (e, e^2)$, this equals $c = 1/\psi(h)$, and if $c \in [e^2, \infty)$, this is less than or equal to $\lfloor \log(c) \rfloor \sqrt{c} < c = 1/\psi(h)$. Thus, the expected number of iterations for sampling $\tilde{V}$ is less than or equal to $1/\psi(h)$. If $c \in (e, \infty)$, the upper bound $\lfloor \log(c) \rfloor c^{1/\lfloor \log(c) \rfloor}$ for the expected number of iterations is bounded from above by $c^{1/\lfloor \log(c) \rfloor} \log(c)$. With $c^{1/\lfloor \log(c) \rfloor} = e^{\log(c)/\lfloor \log(c) \rfloor} \leq e^{(\lfloor \log(c) \rfloor + 1)/\lfloor \log(c) \rfloor} \leq e^2$ it follows that the expected number of iterations is bounded from above by $e^2 \log(c)$. Thus, the complexity $\mathcal{O}\big(\log(1/\psi(h))\big)$ is proven. □

The functions $\psi^{1/m}$ being in Ψ_∞ for all $m \in \mathbb{N}$ is a rather weak assumption, e.g., all generators listed in Table 2.1 share this property. The *fast rejection algorithm* for sampling the distribution $\tilde{F} = \mathcal{LS}^{-1}[\tilde{\psi}]$ with the strategy presented in Theorem 4.2.3 is given as follows. Note that the complexity $\mathcal{O}(1/\psi(h))$ of the standard rejection algorithm is reduced to $\mathcal{O}(\log(1/\psi(h)))$ for any ψ and h, such that $-\log(\psi(h)) > 1$.

Algorithm 4.2.4 (Fast rejection)
Let $\psi \in \Psi_\infty$, such that $\psi^{1/m} \in \Psi_\infty$ for all $m \in \mathbb{N}$, and let $\tilde{\psi}(t) := \psi(h+t)/\psi(h)$, $t \in [0, \infty]$, where $h \in [0, \infty)$. For m as in Equation (4.3), sample i.i.d. $\tilde{V}_k \sim \tilde{F}_m = \mathcal{LS}^{-1}[\tilde{\psi}^{1/m}]$, $k \in \{1, \ldots, m\}$, with the standard rejection algorithm and return $\tilde{V} = \sum_{k=1}^m \tilde{V}_k$.

The fast rejection algorithm will play an important role in the sequel for efficiently sampling exponentially tilted stable distributions. As we will see, these distributions will appear at several points in what follows, e.g., for sampling nested Clayton copulas.

4.2.2 Sampling exponentially tilted stable distributions

Tilting the Clayton generator $\psi(t) = (1+t)^{-1/\vartheta}$ with $h \in [0, \infty)$ leads to $\tilde{\psi}(t) = \psi(h+t)/\psi(h) = (1+t/(1+h))^{-1/\vartheta}$ which generates the same Clayton copula as ψ for any $h \in [0, \infty)$. Hence, tilting a Clayton generator does not lead to an interesting tilted Archimedean generator. However, if we start with the Gumbel generator $\psi(t) = \exp(-V_0 t^\alpha)$ for $\alpha \in (0, 1]$ and $V_0 \in (0, \infty)$, corresponding to an $S(\alpha, 1, (\cos(\alpha \pi/2) V_0)^{1/\alpha}, V_0 \mathbb{1}_{\{\alpha = 1\}}; 1)$ distribution, the resulting tilted Gumbel generator, given by

$$\tilde{\psi}(t) = \exp\big(-V_0((h+t)^\alpha - h^\alpha)\big), \ t \in [0, \infty], \tag{4.4}$$

properly generalizes the Archimedean family of Gumbel. Distributions obtained by exponentially tilting stable distributions are referred to as *exponentially tilted stable distributions* with *tilt h* and we denote the distribution with Laplace-Stieltjes transform as given in (4.4) by

$$\tilde{S}(\alpha, 1, (\cos(\alpha\pi/2)V_0)^{1/\alpha}, V_0\mathbb{1}_{\{\alpha=1\}}, h\mathbb{1}_{\{\alpha\neq 1\}}; 1). \tag{4.5}$$

Exponentially tilted stable distributions first appeared in Tweedie (1984) and Hougaard (1986). They are also considered by Brix (1999). For applications, see, e.g., McCulloch (2003) and McCulloch and Lee (2007) in the context of option pricing. For a more recent treatment, see Barndorff-Nielson and Shephard (2001), who also deal with sampling algorithms for these distributions. Their proposed method originates from Rosiński (2001). This algorithm is also presented in Schoutens (2003). Rosiński (2007) dedicates a whole paper to exponentially tilted stable distributions. Ridout (2008) examines sampling algorithms based on numerical inversion of Laplace transforms and also considers exponentially tilted stable distributions. His study includes the algorithm of Rosiński (2007).

In this subsection, we briefly recall these known approaches and present a new strategy for sampling the exponentially tilted stable distribution as given in (4.5) with corresponding Laplace-Stieltjes transform as in (4.4). Sampling such random variates will be required for a generator transform addressed in Subsection 4.2.3. Note that we have already encountered exponentially tilted Archimedean generators in Subsection 3.3.6 when sampling a nested Clayton copula. Further, the Archimedean families numbered 13 and 19 also lead to a ψ_{01} corresponding to an exponentially tilted stable distribution. Moreover, the family of Gumbel and the one numbered 12 can be obtained as special cases. In Subsection 3.3.6 it turned out that the investigated algorithms for numerical inversion of Laplace transforms had difficulties in accessing (4.5). We will therefore take a closer look at sampling approaches for this distribution and specifically compare them in the case of a nested Clayton copula in Subsection 4.4.3.

Rosiński (2007) considers series representations of Lévy processes, including those with exponentially tilted stable distributed increments, and derives approximations for simulations. For sampling

$$\tilde{S}' \sim \tilde{S}(\alpha, 1, (\cos(\alpha\pi/2)V_0)^{1/\alpha}h, V_0h^\alpha\mathbb{1}_{\{\alpha=1\}}, \mathbb{1}_{\{\alpha\neq 1\}}; 1) \tag{4.6}$$

with corresponding Laplace-Stieltjes transform $\exp\bigl(-V_0h^\alpha((1+t)^\alpha-1)\bigr)$, Rosiński (2007) derives the stochastic representation

$$\tilde{S}' = \sum_{j=1}^{\infty} \min\left\{\left(\frac{V_0}{\Gamma(1-\alpha)\gamma_j}\right)^{1/\alpha} h, e_j u_j^{1/\alpha}\right\}, \tag{4.7}$$

4 Directly sampling nested Archimedean copulas

where $(u_j)_{j\in\mathbb{N}}$ is a sequence of independent and identically distributed U[0,1] random variables, $(e_j)_{j\in\mathbb{N}}$, $(\tilde{e}_j)_{j\in\mathbb{N}}$ are sequences of independent and identically distributed Exp(1) random variables, where all sequences are assumed to be independent, and $\gamma_j := \sum_{i=1}^{j} \tilde{e}_i$, $j \in \mathbb{N}$. The random variable $\tilde{S} := \tilde{S}'/h$ then follows the exponentially tilted stable distribution as given in (4.5). Rosiński (2007) also obtains a stochastic representation for a Lévy process $(\tilde{S}'_t)_{t\in[0,T]}$, $T > 0$, with increments of unit length being distributed as in (4.6), given by

$$\tilde{S}'_t = \sum_{j=1}^{\infty} \min\left\{\left(\frac{TV_0}{\Gamma(1-\alpha)\gamma_j}\right)^{1/\alpha} h, e_j u_j^{1/\alpha}\right\} \mathbb{1}_{(0,t/T]}(\tilde{u}_j), \quad (4.8)$$

where $(\tilde{u}_j)_{j\in\mathbb{N}}$ is a sequence of independent and identically distributed U[0,1] random variables, independent of the other sequences of random variables. The resulting algorithm for sampling n random variates which approximately follow the distribution in (4.5) is given as follows, see also Ridout (2008).

Algorithm 4.2.5 (Rosiński (2007))

(1) Choose a truncation point $J \in \mathbb{N}$ for the sum in (4.8) and the number n of random variates to be generated.

(2) Generate independent sequences $(u_j)_{j=1}^J$, $(\tilde{u}_j)_{j=1}^J$, $(e_j)_{j=1}^J$, and $(\tilde{e}_j)_{j=1}^J$ with independent $u_j, \tilde{u}_j \sim$ U[0,1] and $e_j, \tilde{e}_j \sim$ Exp(1) for all $j \in \{1, \ldots, J\}$.

(3) For $t \in \{0, \ldots, n\}$, compute $\tilde{S}'_t$ as in Representation (4.8), where the sum is truncated at J.

(4) Return $(S_1, \ldots, S_n)$, where $S_t := (\tilde{S}'_t - \tilde{S}'_{t-1})/h$, $t \in \{1, \ldots, n\}$.

Similarly to our approach given in Chapter 3, Ridout (2008) considers numerical inversion of Laplace transforms for sampling purposes, including exponentially tilted stable distributions. He compares his numerical algorithm, called rlaptrans, implemented in the statistical software package R, to the algorithm of Rosiński (2007) for the parameters $(\alpha, V_0, h) \in \{(0.5, 1/\sqrt{2}, 0.5), (0.75, 4^{3/4}/3, 0.25)\}$ and also investigates different choices of truncation points $J \in \{10\,000, 25\,000, 50\,000, 100\,000\}$. Ridout (2008) notes that the algorithm of Rosiński (2007) turns out to be slower (even for $J = 10\,000$) and biased (even for $J = 100\,000$) in comparison to his algorithm rlaptrans. He additionally investigates another numerical sampling algorithm, called rtweedie, based on a combination of a series representation and a Fourier inversion technique for the exponentially tilted stable density. Moreover, he considers the rejection algorithm rdevroye based on an envelope found by using characteristic functions, see Devroye (1986, p. 696), adapted to the setup of non-negative random variables. Based on the run time,

4.2 Nested Archimedean copulas based on generator transformations

Ridout (2008) notes that these methods can not compete with `rlaptrans` for the examined parameter choices.

The problem of sampling an exponentially tilted stable distribution is even more demanding in the case of sampling a generator ψ_{01} of Type (4.4) involved in a nested Archimedean copula. The reason for this is that V_0 acts as an additional parameter on the exponentially tilted stable distribution which possibly changes with every vector of random variates to be generated. Thus, one has to sample vectors of random variates from different exponentially tilted stable distributions rather than from a distribution with fixed parameters. For a nested Clayton copula, e.g., V_0 is a gamma distributed random variable and thus almost surely changes for every vector of random variates to be generated. Algorithms requiring numerically demanding setup steps for sampling F_{01} may therefore be inadequate for applications in large-scale simulation studies. For this reason, Algorithm 4.2.5 of Rosiński (2007) can not be efficiently applied. Instead, in order to use the idea of Rosiński (2007) for sampling V_{01} following the distribution given in (4.5), one has to truncate the series in Representation (4.7) at some large $J \in \mathbb{N}$. This requires generating $3J$ random variates for each random variate V_{01} to be drawn, which is demanding. Moreover, this is only an approximate sampling algorithm and the error is hard to control.

In contrast to the approaches of Rosiński (2007) and Ridout (2008), the standard rejection algorithm is exact and does not rely on approximations. The only problem is its complexity. If ψ_{01} takes the form as given in (4.4), the complexity is $\mathcal{O}(e^{V_0 h^\alpha})$, i.e., exponential in $V_0 h^\alpha$. For sampling a nested Clayton copula, this becomes a problem if ϑ_0 is small, since $\mathbb{E}[V_0]$ is large in this case.

To summarize, all known algorithms for sampling exponentially tilted stable distributions appearing in nested Archimedean copulas are either inefficient or the numerical errors due to approximations are hard to control. However, the fast rejection algorithm efficiently solves this problem. Since $\tilde{\psi}$ as given in (4.4) is a tilted generator based on $\psi(t) = \exp(-V_0 t^\alpha)$ with tilt h, the complexity of the fast rejection algorithm is $\mathcal{O}(V_0 h^\alpha)$, i.e., only linear in $V_0 h^\alpha$ instead of exponential, for every V_0 such that $V_0 h^\alpha > 1$. Further, the fast rejection algorithm is exact and does not require time-consuming setup steps. We will see its good performance in comparison to the other presented algorithms in Subsection 4.4.3 when we sample nested Clayton copulas.

Let us close this subsection with the remark that a result similar to the fast rejection algorithm for sampling the distribution given in (4.5) can be obtained by writing $\tilde{\psi}$ in (4.4) as $\tilde{\psi} = \psi_2 \psi_3^{\lfloor V_0 h^\alpha \rfloor}$ with $\psi_2(t) = \exp\bigl(-(V_0 h^\alpha - \lfloor V_0 h^\alpha \rfloor)((1+t/h)^\alpha - 1)\bigr)$ and $\psi_3(t) = \exp(-((1+t/h)^\alpha - 1))$. One can therefore sample an exponentially tilted stable random variable $\tilde{S}$ following the distribution given in (4.5) as $\tilde{S} := S_2 + \sum_{k=1}^{\lfloor V_0 h^\alpha \rfloor} S_{3k}$, where S_2 is a random variable with distribution corresponding to

4 Directly sampling nested Archimedean copulas

ψ_2, independently of S_{3k}, $k \in \{1, \ldots, \lfloor V_0 h^\alpha \rfloor\}$, which are independent and identically distributed random variables following the distribution with Laplace-Stieltjes transform ψ_3. Note that the generators ψ_j, $j \in \{2, 3\}$, are of type $\tilde{\psi}_j(t/h)$. Therefore, the distribution corresponding to ψ_j can be sampled by sampling the one corresponding to $\tilde{\psi}_j(t)$ and dividing the generated random variates by h. If we apply the standard rejection algorithm to sample the random variates involved, the complexity of this sampling strategy is of order $e^{V_0 h^\alpha - \lfloor V_0 h^\alpha \rfloor} + \lfloor V_0 h^\alpha \rfloor e$. By writing x for $V_0 h^\alpha$, we may investigate the performance of this algorithm for all $V_0 \in (0, \infty)$ and $h \in [0, \infty)$ by analyzing the function $\exp(x - \lfloor x \rfloor) + \lfloor x \rfloor e$ for all $x \in [0, \infty)$. Recall that the complexity of the fast rejection algorithm for sampling the distribution corresponding to $\tilde{\psi}$ in terms of $x = V_0 h^\alpha$ is of order $\exp(x)$ if $x \in [0, 1]$ and $\min\{\lfloor x \rfloor \exp(x/\lfloor x \rfloor), \lceil x \rceil \exp(x/\lceil x \rceil)\}$ if $x \in (1, \infty)$. Comparing the complexities as functions of $x \in [0, \infty)$ leads to a slightly worse complexity for the algorithm based on ψ_2 and ψ_3 than for the fast rejection algorithm. However, ψ_3 is now independent of V_0. By considering $\tilde{\psi}_3$, it is even independent of h. Therefore, algorithms requiring setup steps for efficiently sampling the distribution corresponding to $\tilde{\psi}_3$ might be applied. Further, numerical inversion of Laplace transforms is another option since numerical inversion of $\tilde{\psi}_3$ is considerably less critical than that of $\tilde{\psi}$. This is due to the fact that the former only depends on the parameter α but neither on V_0 nor h. Thus, careful checks can be made to find optimal parameters for the numerical inversion procedure under consideration beforehand.

4.2.3 A general nesting transformation

A transformation applied to an Archimedean generator is particularly useful if the resulting transformed generator allows for nesting. In this subsection, we introduce such a transformation. We precisely characterize the outer and inner distribution functions corresponding to the transformed generators by giving stochastic representations. These stochastic representations allow for sampling. Hence, the resulting nested Archimedean copulas can also be sampled efficiently as long as one can sample the distribution corresponding to the base generator which is originally transformed.

The Archimedean generator transform we present is motivated by the following observation. Both a nested Gumbel copula and a nested Archimedean copula based on the family numbered 12 share the same functional form of the node $\psi_0^{-1} \circ \psi_1$, given by $\psi_0^{-1}(\psi_1(t)) = t^{\vartheta_0/\vartheta_1}$. Similarly, both the nodes of a nested Clayton copula and a nested Archimedean copula based on the family numbered 13 share the same functional form, given by $\psi_0^{-1}(\psi_1(t)) = (1+t)^{\vartheta_0/\vartheta_1} - 1$. A general transformation which subsumes these two cases and many others is given as follows.

4.2 Nested Archimedean copulas based on generator transformations

Theorem 4.2.6
Let $\psi \in \Psi_\infty$ and $c \in [0, \infty)$.
(1) If $\tilde{\psi}(t) := \psi((c^\vartheta + t)^{1/\vartheta} - c)$, $t \in [0, \infty]$, with $\vartheta \in [1, \infty)$, then $\tilde{\psi} \in \Psi_\infty$ and
$$\tilde{V} := \tilde{S}V^\vartheta \sim \tilde{F} := \mathcal{LS}^{-1}[\tilde{\psi}],$$
where $V \sim F := \mathcal{LS}^{-1}[\psi]$ and $\tilde{S} \sim \mathrm{S}(1/\vartheta, 1, \cos^\vartheta(\pi/(2\vartheta)), \mathbb{1}_{\{\vartheta=1\}}, (cV)^\vartheta \mathbb{1}_{\{\vartheta \neq 1\}}; 1)$.
(2) If $\psi_j(t) := \psi((c^{\vartheta_j} + t)^{1/\vartheta_j} - c)$, $t \in [0, \infty]$, with $\vartheta_j \in [1, \infty)$, $j \in \{0, 1\}$, and $\vartheta_0 \leq \vartheta_1$, then $(\psi_0^{-1} \circ \psi_1)'$ is completely monotone and $\psi_{01}(t; V_0)$ takes the form as given in (4.4) with $\alpha := \vartheta_0/\vartheta_1$ and $h := c^{\vartheta_1}$. Thus, F_{01} is an exponentially tilted stable distribution as given in (4.5).

Proof
First consider (1). Since $(c^\vartheta + t)^{1/\vartheta} - c \geq 0$ has a completely monotone derivative, it follows from Proposition 2.1.5 (2) that $\tilde{\psi} \in \Psi_\infty$. Let f_S denote the density of $\mathrm{S}(1/\vartheta, 1, \cos^\vartheta(\pi/(2\vartheta)), \mathbb{1}_{\{\vartheta=1\}}; 1)$ with Laplace-Stieltjes transform $\psi_S(t) = \exp(-t^{1/\vartheta})$ and $f_{\tilde{S}|V}$ the conditional density of $\tilde{S}$ given V, which is $f_{\tilde{S}|V}(\tilde{s}|V) = \exp(-h\tilde{s})f_S(\tilde{s})/\psi_S(h)$ with $h = (cV)^\vartheta$, i.e., $f_{\tilde{S}|V}(\tilde{s}|V) = \exp(-(cV)^\vartheta \tilde{s} + cV)f_S(\tilde{s})$. Then

$$\mathcal{LS}[F_{\tilde{S}V^\vartheta}](t) = \mathbb{E}[\exp(-t\tilde{S}V^\vartheta)] = \mathbb{E}[\mathbb{E}[\exp(-t\tilde{S}V^\vartheta)|V]]$$
$$= \mathbb{E}\left[\int_0^\infty \exp(-t\tilde{s}V^\vartheta)f_{\tilde{S}|V}(\tilde{s}|V)\,d\tilde{s}\right]$$
$$= \mathbb{E}\left[\exp(cV)\int_0^\infty \exp(-\tilde{s}V^\vartheta(c^\vartheta + t))f_S(\tilde{s})\,d\tilde{s}\right],$$
$$= \mathbb{E}\left[\exp(cV)\exp\left(-(V^\vartheta(c^\vartheta + t))^{1/\vartheta}\right)\right],$$
$$= \mathbb{E}\left[\exp\left(-V((c^\vartheta + t)^{1/\vartheta} - c)\right)\right] = \psi((c^\vartheta + t)^{1/\vartheta} - c) = \tilde{\psi}(t).$$

Thus, $\tilde{V} \sim \tilde{F}$. For (2), note that $\psi_0^{-1}(\psi_1(t)) = (h+t)^\alpha - h^\alpha$, thus $(\psi_0^{-1} \circ \psi_1)'$ is completely monotone. Further, it is easily verified that $\psi_{01}(t; V_0)$ takes the form as given in (4.4) with $\alpha := \vartheta_0/\vartheta_1$ and $h := c^{\vartheta_1}$, so F_{01} is an exponentially tilted stable distribution as given in (4.5). □

Remark 4.2.7
Theorem 4.2.6 encompasses several known nested Archimedean copulas as special cases. As addressed above, sampling nested Clayton ($c = 1$) and Gumbel ($c = 0$) copulas, as well as nested Archimedean copulas based on the families numbered 12 ($c = 0$) and 13 ($c = 0$) follow from this result. Further, nested Archimedean copulas based on the family numbered 19 ($c = e$) can be obtained as special cases. Moreover, all nested outer power Archimedean copulas follow from Theorem 4.2.6 ($c = 0$).

4 Directly sampling nested Archimedean copulas

Theorem 4.2.6 allows us to start with any Archimedean generator $\psi \in \Psi_\infty$ and build a nested Archimedean copula based on the Archimedean family $\psi((c^\vartheta + t)^{1/\vartheta} - c)$. The following result summarizes the dependence properties of this transformation. Note that we obtained formulas for Kendall's tau for the families numbered 12, 13, and 14 in Table 2.2 by applying the first part of this result.

Proposition 4.2.8

Let $\psi \in \Psi_\infty$, $c \in [0, \infty)$, and $\tilde{\psi}(t) := \psi((c^\vartheta + t)^{1/\vartheta} - c)$, $t \in [0, \infty]$, with $\vartheta \in [1, \infty)$. Further, let τ, λ_l, and λ_u denote Kendall's tau and the lower and upper tail-dependence coefficients, respectively, corresponding to the Archimedean copula generated by ψ, where we assume that the latter two exist. Then,

(1) Kendall's tau $\tilde{\tau}$ corresponding to the Archimedean copula generated by $\tilde{\psi}$ is given by

$$\tilde{\tau} = 1 - \frac{4}{\vartheta} \int_0^\infty (1 + c/t)(1 - 1/(1 + t/c)^\vartheta) t (\psi'(t))^2 \, dt \qquad (4.9)$$

and $\tilde{\tau} \geq \tau$. If $c = 0$ then $\tilde{\tau} = 1 - (1-\tau)/\vartheta$.

(2) If the lower and upper tail-dependence coefficients $\tilde{\lambda}_l$ and $\tilde{\lambda}_u$ corresponding to the Archimedean copula generated by $\tilde{\psi}$ exist, then $\tilde{\lambda}_l \geq \lambda_l$ and $\tilde{\lambda}_u \geq \lambda_u$.

Proof

For (1), apply the second equation in Formula (2.10) and substitute $(c^\vartheta + t)^{1/\vartheta} - c$ to obtain Identity (4.9). Using the same argument leads to the last statement of (1). Now consider the inequality $\tilde{\tau} \geq \tau$. By Identity (4.9) it remains to show that $(1 + c/t)(1 - 1/(1 + t/c)^\vartheta) \leq \vartheta$. This can be achieved by verifying that the derivative of the left-hand side is monotonically decreasing in t with limit equal to ϑ as $t \searrow 0$. For (2), verify that $f(t; c, \vartheta) := (c^\vartheta + t)^{1/\vartheta} - c$ is concave and that $f(0; c, \vartheta) = 0$. Thus, $f(2t; c, \vartheta) \leq 2f(t; c, \vartheta)$. This implies $\tilde{\psi}(2t)/\tilde{\psi}(t) \geq \psi(2f(t;c,\vartheta))/\psi(f(t;c,\vartheta))$. Since $f(t; c, \vartheta) \to \infty$ as $t \to \infty$, $\tilde{\lambda}_l \geq \lambda_l$ follows. The same argument can be applied to show that $\tilde{\lambda}_u \geq \lambda_u$. □

If we know how to sample $F = \mathcal{LS}^{-1}[\psi]$, we do so for the nested Archimedean copula based on the transformed generator $\tilde{\psi}(t) = \psi((c^\vartheta + t)^{1/\vartheta} - c)$. The exponentially tilted stable distributions involved in Theorem 4.2.6 can be sampled with the fast rejection algorithm. The following recursive algorithm gives instructions for sampling the fully-nested Archimedean copula of Type (2.4), constructed based on the generator transformation $\tilde{\psi}$.

Algorithm 4.2.9

Let $\psi \in \Psi_\infty$ and $c \in [0, \infty)$. Further, let $\psi_i(t) := \psi((c^{\vartheta_i} + t)^{1/\vartheta_i} - c)$, $i \in \{0, \ldots, d-2\}$, $d \geq 2$, with $1 \leq \vartheta_0 \leq \cdots \leq \vartheta_{d-2} < \infty$.

(1) Sample $V_0 \sim F_0 := \mathcal{LS}^{-1}[\psi_0]$ via $V_0 := \tilde{S}V^{\vartheta_0}$, where $V \sim F := \mathcal{LS}^{-1}[\psi]$ and $\tilde{S} \sim \tilde{S}(1/\vartheta_0, 1, \cos^{\vartheta_0}(\pi/(2\vartheta_0)), \mathbb{1}_{\{\vartheta_0=1\}}, (cV)^{\vartheta_0}\mathbb{1}_{\{\vartheta_0 \neq 1\}}; 1)$.

(2) Sample $X_1 \sim U[0,1]$ independently of V_0 and $(X_2, \ldots, X_d)^T \sim C(u_2, \ldots, u_d; \psi_{01}(\cdot; V_0), \ldots, \psi_{0d-2}(\cdot; V_0))$ independently of X_1, where for $i \in \{1, \ldots, d-2\}$, $\psi_{0i}(t; V_0) := \exp(-V_0((h_i + t)^{\alpha_i} - h_i^{\alpha_i}))$, $h_i := c^{\vartheta_i}$, $\alpha_i := \vartheta_0/\vartheta_i$, corresponds to an $\tilde{S}(\alpha_i, 1, (\cos(\alpha_i\pi/2)V_0)^{1/\alpha_i}, V_0\mathbb{1}_{\{\alpha_i=1\}}, h_i\mathbb{1}_{\{\alpha_i \neq 1\}}; 1)$ distribution.

(3) Return $\boldsymbol{U}$, where $U_j := \psi_0(-\log(X_j)/V_0)$, $j \in \{1, \ldots, d\}$.

4.3 Nested Archimedean copulas based on different Archimedean families

A way to bring in two Archimedean generators for constructing a new one is the following. Given Archimedean generators $\psi, \psi_0 \in \Psi_\infty$ such that $\psi^\alpha \in \Psi_\infty$ for all $\alpha \in (0, \infty)$, the function

$$\tilde{\psi} := \psi_0 \circ (-\log) \circ \psi \qquad (4.10)$$

defines an Archimedean generator in Ψ_∞, see Proposition 2.1.5 (2), (5). As mentioned above, all generators ψ listed in Table 2.1 satisfy $\psi^\alpha \in \Psi_\infty$ for all $\alpha \in (0, \infty)$.

Concerning the dependence properties of the Archimedean copula generated by $\tilde{\psi}$, note that the Cauchy-Schwarz inequality implies that $\psi(t) \leq \sqrt{\psi(2t)}$. This in turn implies that $-\log(\psi(2t)) \leq -2\log(\psi(t))$. Assuming that the limits involved exist, it follows from the first equalities in Formulas (2.11) and (2.12) that

$$\tilde{\lambda}_l \geq \lambda_l \text{ and } \tilde{\lambda}_u \geq \lambda_u.$$

By writing $\tilde{\psi}$ as

$$\tilde{\psi}(t) = \int_0^\infty \psi^x(t) \, dF_0(x),$$

where $F_0 := \mathcal{LS}^{-1}[\psi_0]$, we see that $\tilde{\psi}$ is simply the generator ψ with randomized power x according to the mixing distribution F_0. The following algorithm uses this construction principle to sample the distribution $\tilde{F}$ corresponding to $\tilde{\psi}$.

Algorithm 4.3.1

(1) Sample $V_0 \sim F_0 = \mathcal{LS}^{-1}[\psi_0]$.

4 Directly sampling nested Archimedean copulas

(2) Sample and return $\tilde{V} \sim \mathcal{LS}^{-1}[\psi^{V_0}]$.

By a similar reasoning one also sees that $\tilde{\psi} := \psi_0 \circ \psi_1^{-1} \circ \psi$ is a proper Archimedean generator for any two generators $\psi, \psi_1 \in \Psi_\infty$ that satisfy the sufficient nesting condition. Note that $\tilde{\psi}$ defined this way even makes sense for all generators $\psi \in \Psi_\infty$ as long as $\psi_0 \circ \psi_1^{-1}$ is absolutely monotone on $[0, 1]$, see Proposition 2.1.5 (4).

Except when F_0 is discrete on $\mathbb{N}$, there is no general strategy known for accessing Algorithm 4.3.1 (2) when the distribution corresponding to ψ is easy to sample. However, considering the specific form of a given generator ψ, one usually finds an algorithm for sampling the distribution corresponding to ψ^{V_0}. For example, by taking $\psi_0(t) = 1/(1+t)$, i.e., Clayton's generator with parameter 1, and $\psi(t) = (1 + t/e^\vartheta)^{-1/\vartheta}$, i.e., a scaled Clayton generator with corresponding distribution $\Gamma(1/\vartheta, e^\vartheta)$, $\tilde{\psi}$ is easily seen to be the Archimedean generator numbered 19. Applying Algorithm 4.3.1, a random variable $\tilde{V} \sim \mathcal{LS}^{-1}[\tilde{\psi}]$ can be constructed by taking $V_0 \sim \text{Exp}(1)$ and putting $\tilde{V} := \Gamma(V_0/\vartheta, e^\vartheta)$. By taking $\psi_0(t) = (1+t)^{-1/\vartheta}$, i.e., Clayton's generator, and $\psi(t) = 1/(1 + t/e)$, i.e., a scaled Clayton generator with parameter 1 with corresponding distribution $\text{Exp}(e)$, $\tilde{\psi}$ is easily seen to be the Archimedean generator numbered 20. Similarly to the family numbered 19, one obtains the stochastic representation $\Gamma(\Gamma(1/\vartheta, 1), e)$ for a random variable following $\mathcal{LS}^{-1}[\tilde{\psi}]$. We have therefore derived the remaining entries in Table 2.1.

Now let us apply Construction (4.10) to nested Archimedean copulas. Not every combination of generators $\psi_0, \psi_1 \in \Psi_\infty$ is known to lead to a valid nested Archimedean copula. If generators belonging to different Archimedean families are involved, the sufficient nesting condition does not necessarily hold. For example, considering the Archimedean generators listed in Table 2.1 with parameter 0.5 for Ali-Mikhail-Haq's family and 1.2 for all others, one can infer from the second derivative of $\psi_0^{-1} \circ \psi_1$ that the sufficient nesting condition is already violated by 60 out of the 90 possible family combinations involving different Archimedean families. Further, if ψ_0 and ψ_1 are generators for Clayton's and Gumbel's copula, respectively, calculating the first derivative shows that $(\psi_0^{-1} \circ \psi_1)'$ is not completely monotone for any parameter choice. Moreover, $(\psi_1^{-1} \circ \psi_0)'$ is completely monotone only if the parameter of Gumbel's copula equals one, i.e., if ψ_1 generates the independence copula.

However, by choosing $\psi_1(t) := \psi_0\bigl(-\log(\psi(t))\bigr)$ for $\psi, \psi_0 \in \Psi_\infty$ such that $\psi^\alpha \in \Psi_\infty$ for all $\alpha \in (0, \infty)$, ψ_0 and ψ_1 always satisfy the sufficient nesting condition. We have already seen that $\psi_1 \in \Psi_\infty$. Showing that $\psi_0^{-1} \circ \psi_1$ has completely monotone derivative boils down to showing that $-\log \psi$ has completely monotone derivative, which holds by Proposition 2.1.5 (5).

4.3 Nested Archimedean copulas based on different Archimedean families

This choice for ψ_1 implies that the inner generator takes the form

$$\psi_{01}(t; V_0) = \psi^{V_0}(t),$$

which means that ψ_1 is chosen such that $\psi_{01}(t; V_0)$ is a power in ψ, randomized according to the distribution corresponding to ψ_0. Starting with a generator ψ such that $\psi^\alpha \in \Psi_\infty$ for all $\alpha \in (0, \infty)$ and whose powers correspond to distributions that are easy to sample, one may thus build and sample the corresponding nested Archimedean copula. This gives rise to a simple approach for constructing nested Archimedean copulas. By starting with some generator ψ_0 with known $F_0 := \mathcal{LS}^{-1}[\psi_0]$, one chooses ψ_1 such that ψ_{01} equals some Archimedean generator with known corresponding distribution function F_{01}. Thus, the resulting structure can easily be sampled.

The idea of randomizing generator powers can be applied to sample nested Archimedean copulas constructed via generators belonging to different Archimedean families. In Subsection 3.3.7 we originally applied a procedure for numerically inverting Laplace transforms to sample a fully-nested A(C) copula, i.e., a Clayton copula nested into an Ali-Mikhail-Haq copula. With the help of tilted Archimedean generators, the generator transformation addressed in Theorem 4.2.6, and randomized generator powers, we are now able to find explicit sampling strategies for several combinations of different Archimedean families for which the sufficient nesting condition holds.

Theorem 4.3.2

(1) Let ψ_0 be Ali-Mikhail-Haq's generator with parameter ϑ_0 and let ψ_1 be Clayton's generator with parameter ϑ_1. Then, the sufficient nesting condition holds for all $\vartheta_1 \in [1, \infty)$. Further,

$$V_{01} = \tilde{S} V^{\vartheta_1},$$

where

$$V \sim \Gamma(V_0, 1/(1-\vartheta_0)), \ \tilde{S} \sim \tilde{S}(1/\vartheta_1, 1, \cos^{\vartheta_1}(\pi/(2\vartheta_1)), \mathbb{1}_{\{\vartheta_1=1\}}, V^{\vartheta_1} \mathbb{1}_{\{\vartheta_1 \neq 1\}}; 1).$$

(2) Let ψ_0 be Ali-Mikhail-Haq's generator with parameter ϑ_0 and let ψ_1 be the generator of the family numbered 19 with parameter ϑ_1. Then, the sufficient nesting condition holds for all ϑ_0 and ϑ_1. Further, $V_{01} \sim F_{01}$ can be sampled with the fast rejection algorithm, since $\psi_{01}(t; V_0) = \psi(h+t)/\psi(h)$ with $h := e^{\vartheta_1} - 1$, $c = 1/\vartheta_0^{V_0}$, and

$$F_m := \mathcal{LS}^{-1}[\psi^{1/m}] = \Gamma((1-\vartheta_0)\Gamma(V_0/m, 1)/(\vartheta_0 \vartheta_1), 1), \ m \in \mathbb{N}.$$

4 Directly sampling nested Archimedean copulas

For the family combination A(20), proceed similarly as for A(19), with $\vartheta_1 \in [1, \infty)$, where $h := e - 1$, $c = 1/\vartheta_0^{V_0}$, and

$$F_m := \mathcal{LS}^{-1}[\psi^{1/m}] = \Gamma((1/\vartheta_0 - 1)^{\vartheta_1} S\Gamma(V_0/m, 1)^{\vartheta_1}, 1), \ m \in \mathbb{N},$$

with $S \sim \mathrm{S}(1/\vartheta_1, 1, \cos^{\vartheta_1}(\pi/(2\vartheta_1)), \mathbb{1}_{\{\vartheta_1=1\}}; 1)$.

(3) Let ψ_0 be Clayton's generator with parameter ϑ_0 and let ψ_1 be the generator of the family numbered 12 with parameter ϑ_1. Then, the sufficient nesting condition holds for all $\vartheta_0 \in (0, 1]$. Further,

$$V_{01} = S\tilde{S}^{\vartheta_1},$$

where

$$S \sim \mathrm{S}(1/\vartheta_1, 1, \cos^{\vartheta_1}(\pi/(2\vartheta_1)), \mathbb{1}_{\{\vartheta_1=1\}}; 1),$$
$$\tilde{S} \sim \tilde{\mathrm{S}}(\vartheta_0, 1, (\cos(\pi\vartheta_0/2)V_0)^{1/\vartheta_0}, V_0 \mathbb{1}_{\{\vartheta_0=1\}}, \mathbb{1}_{\{\vartheta_0 \neq 1\}}; 1).$$

For the family combination C(14) with $\vartheta_0 \vartheta_1 \in (0, 1]$, apply the procedure for C(12) with ϑ_0 replaced by $\vartheta_0 \vartheta_1$.

(4) Let ψ_0 be Clayton's generator with parameter ϑ_0 and let ψ_1 be the generator of the family numbered 19 with parameter ϑ_1. Then, the sufficient nesting condition holds for all $\vartheta_0 \in (0, 1]$. Further, $V_{01} \sim F_{01}$ can be sampled with the fast rejection algorithm, since $\psi_{01}(t; V_0) = \psi(h+t)/\psi(h)$ with $h := e^{\vartheta_1} - 1$, $c = e^{V_0}$, and

$$F_m := \mathcal{LS}^{-1}[\psi^{1/m}] = \Gamma(S_m, 1), \ m \in \mathbb{N},$$

where $S_m \sim \mathrm{S}(\vartheta_0, 1, (\cos(\pi\vartheta_0/2)V_0/m)^{1/\vartheta_0}/\vartheta_1, V_0 \mathbb{1}_{\{\vartheta_0=1\}}/(\vartheta_1^{\vartheta_0}m); 1)$. For the family combination C(20) with $\vartheta_0 \leq \vartheta_1$, apply the procedure for C(19) with ϑ_1 replaced by 1 and then ϑ_0 replaced by ϑ_0/ϑ_1.

Proof

First consider the claims about the parameter ranges of ϑ_0 and ϑ_1. For (1) and (2), these follow from the fact that the nodes $\psi_0^{-1} \circ \psi_1$ take the form $-\log \psi$ for some ψ for which $\psi^\alpha \in \Psi_\infty$ for all $\alpha \in (0, \infty)$. The corresponding statements then follow from Proposition 2.1.5 (5). For (3) and (4), the claims about the parameter ranges for ϑ_0 and ϑ_1 follow by verifying that $(\psi_0^{-1} \circ \psi_1)'$ is indeed completely monotone, which can be achieved with the tools provided in Proposition 2.1.5.

4.3 Nested Archimedean copulas based on different Archimedean families

Now let us prove the statements about V_{01}. For (1), note that $\psi_{01}(t; V_0) = \psi((1+t)^{1/\vartheta_1} - 1)$, where $\psi(t) = (1 + (1-\vartheta_0)t)^{-V_0}$ corresponds to a $\Gamma(V_0, 1/(1-\vartheta_0))$ distribution. Thus, an application of Theorem 4.2.6 leads to the result as stated.

For (2) and family combination A(19), verify that $\psi_{01}(t; V_0) = \psi(h+t)/\psi(h)$ with $h := e^{\vartheta_1} - 1$ and $\psi := \psi_2 \circ (-\log) \circ \psi_3$, where $\psi_2(t) := (1+t)^{-V_0}$ corresponds to a $\Gamma(V_0, 1)$ distribution and $\psi_3(t) := (1+t)^{-(1-\vartheta_0)/(\vartheta_0\vartheta_1)}$ corresponds to a $\Gamma((1-\vartheta_0)/(\vartheta_0\vartheta_1), 1)$ distribution. The given stochastic representation for $F_m = \mathcal{LS}^{-1}[\psi^{1/m}]$ then follows from Algorithm 4.3.1, since $\psi^\alpha \in \Psi_\infty$ for all $\alpha \in (0, \infty)$. For the family combination A(20), take $h := e - 1$, $\psi_2(t) := (1+t^{1/\vartheta_1})^{-V_0}$, and $\psi_3(t) := (1+t)^{-(1/\vartheta_0-1)^{\vartheta_1}}$, and proceed similarly as before. As ψ_2 is an outer power Clayton generator, it follows from Theorem 4.2.6 that a random variable corresponding to ψ_2 admits the stochastic representation $S\Gamma(V_0, 1)^{\vartheta_1}$ with $S \sim S(1/\vartheta_1, 1, \cos^{\vartheta_1}(\pi/(2\vartheta_1)), \mathbb{1}_{\{\vartheta_1=1\}}; 1)$. Further, ψ_3 corresponds to a $\Gamma((1/\vartheta_0 - 1)^{\vartheta_1}, 1)$ distribution. The given stochastic representation for $F_m = \mathcal{LS}^{-1}[\psi^{1/m}]$ again follows from Algorithm 4.3.1. As before, the fast rejection algorithm may be applied to this setup, with constant c as given.

For (3) and family combination C(12), $\psi_{01}(t; V_0) = \psi(t^{1/\vartheta_1})$, where $\psi(t) := \exp(-V_0((1+t)^{\vartheta_0} - 1))$ corresponds to an $\tilde{S}(\vartheta_0, 1, (\cos(\pi\vartheta_0/2)V_0)^{1/\vartheta_0}, V_0 \mathbb{1}_{\{\vartheta_0=1\}}, \mathbb{1}_{\{\vartheta_0 \neq 1\}}; 1)$ distribution. As ψ_{01} is an outer power family based on the generator ψ, the claim follows from Theorem 4.2.6. The statement for the family combination C(14) is clear.

For (4) and family combination C(19), $\psi_{01}(t; V_0) = \psi(h+t)/\psi(h)$ with $h := e^{\vartheta_1} - 1$ and $\psi := \psi_2 \circ (-\log) \circ \psi_3$, where $\psi_2(t) := \exp(-V_0(t/\vartheta_1)^{\vartheta_0})$ corresponds to a $S(\vartheta_0, 1, (\cos(\pi\vartheta_0/2)V_0)^{1/\vartheta_0}/\vartheta_1, V_0\mathbb{1}_{\{\vartheta_0=1\}}/\vartheta_1^{\vartheta_0}; 1)$ distribution and $\psi_3(t) := 1/(1+t)$ corresponds to an $\text{Exp}(1)$ distribution. The given stochastic representation for $F_m = \mathcal{LS}^{-1}[\psi^{1/m}]$ then follows from Algorithm 4.3.1. Again note that $\psi^\alpha \in \Psi_\infty$ for all $\alpha \in (0, \infty)$, thus the fast rejection algorithm applies. The statement for the family combination C(20) is clear. □

We see from Theorem 4.3.2 that it is possible to have different Archimedean families appearing in a nested Archimedean copula. The resulting structures are interesting in that they allow different marginal copulas to have different tail dependencies. For example, for a nested A(C) copula, all components with marginal Ali-Mikhail-Haq copulas are tail independent, whereas components with marginal Clayton copulas allow for lower tail dependence.

More important than the compatible Archimedean families addressed in Theorem 4.3.2 is the strategy according to which they are built. As we have seen, tilting of Archimedean generators, the generator transformation of Theorem 4.2.6, and generators with randomized powers can be used as building blocks for constructing and sampling nested Archimedean copulas starting with simple Archimedean generators, e.g., the ones listed in Table 2.1. Some of the generators

4 Directly sampling nested Archimedean copulas

listed in Table 2.1 can even be decomposed into simpler pieces. For example, Gumbel's family, an outer power family, can be obtained from Theorem 4.2.6 with $\exp(-t)$, as can the family numbered 13. Further, as we have seen, the distributions corresponding to the generators of the families numbered 19 and 20 can be obtained by randomizing generator powers based on, possibly scaled, Clayton generators. Moreover, the family numbered 12 is an outer power family based on $1/(1+t)$, i.e., Clayton's generator with parameter equal to one. This generator frequently appears as a simple generator to start with, which may answer one of the open questions raised by Nelsen (2005). The presented approaches therefore provide us with a large repertoire of flexible dependence structures.

4.4 Experiments and examples

In this section we consider several experiments and examples. The first two correspond to our findings in Section 4.1. First, we investigate the run times for two different strategies for sampling bivariate Joe copulas for which the distribution associated to the generator is discrete. Then, we present precision and run-time results for different trivariate fully-nested Frank and Joe copulas. Concerning the results of Section 4.2, we compare the algorithms of Rosiński (2007), Ridout (2008), the rejection algorithm proposed by McNeil (2008), and our proposed fast rejection algorithm for generating vectors of random variates from trivariate fully-nested Clayton copulas with different parameters. As an overview, including our findings from Section 4.3, we present precision and run-time results for various nested Archimedean copulas of Type (2.5) with dimensions 1(2) and 50(50). The precision is checked for the former case and run-time results are presented for both cases. Gaussian and t_4 copulas are included as a benchmark. To demonstrate that all algorithms apply to more than two hierarchies and may involve different Archimedean families, we finally sample a four-dimensional fully-nested Archimedean copula with three nesting levels involving three different Archimedean families. All algorithms are implemented and run following the guidelines outlined in Subsection 3.3.1.

4.4.1 Joe copulas

A general approach for generating random variates from a univariate distribution is the inversion method, see Proposition 1.1.7 (2). If F is discrete and the jump locations and heights of F are known, possibly computed beforehand, drawing from F can be achieved by locating the quantiles of generated uniform random variates. In this subsection, we investigate two different

4.4 Experiments and examples

approaches for finding quantiles of the discrete distribution F involved in sampling bivariate Joe copulas with the algorithm of Marshall and Olkin (1988). For each uniform random variate, the direct approach uses a bisection procedure to locate the corresponding quantile. For the second approach, the vector of uniform random variates is sorted first, then the positive integers are searched through monotonically to locate all required quantiles of F, and finally the sorting is undone. At a first glance, each approach can be advantageous over the other depending on the number n of random variates to be drawn from F and the number of precomputed values of F.

Table 4.1 contains the mean run times for these two approaches for different choices of τ and n. For more accurate run-time results, the measurements are based on 1 000 replications, including precomputing the function values of F. The second column either lists the smallest number $x_{\max} < 500\,000$ for which $F(x_{\max}) \geq 1 - \varepsilon = 1 - 10^{-5}$ or $F(x_{\max})$, with $x_{\max} = 500\,000$, if this level is not reached within the first 500 000 positive integers. Sampling is then done via truncation, where we return $x_{\max} + 1$ as U-quantile if $U > F(x_{\max})$, see Subsection 3.3.1. The parameters of the sampled Joe copulas are chosen such that the resulting Kendall's tau cover an adequate range for most applications. The values 0.01 and 0.05 are chosen to reflect a small number of precomputed values of F.

		$n = 1\,000$		$n = 10\,000$		$n = 100\,000$	
τ	$x_{\max}/F$	direct	sort	direct	sort	direct	sort
0.01	1 973	0.0012	0.0012	0.0078	0.0086	0.0744	0.0839
0.05	20 049	0.0056	0.0056	0.0123	0.0129	0.0798	0.0888
0.1	117 106	0.0288	0.0289	0.0356	0.0364	0.1015	0.1111
0.2	0.99996	0.1186	0.1184	0.1251	0.1267	0.1927	0.2055
0.3	0.99970	0.1182	0.1196	0.1253	0.1273	0.1929	0.2030
0.4	0.99833	0.1184	0.1203	0.1250	0.1281	0.1935	0.2041
0.5	0.99270	0.1181	0.1200	0.1252	0.1271	0.1945	0.2045
0.6	0.97388	0.1184	0.1200	0.1252	0.1272	0.1947	0.2029

Table 4.1 Mean run times in seconds for sampling bivariate Joe copulas with two different algorithms for finding quantiles.

For all investigated sample sizes n and dependencies τ, there is no significant improvement in speed when using one approach in favor of the other. The overall small run times show that Joe copulas can be sampled efficiently with the Marshall-Olkin algorithm. Sampling such copulas

4 Directly sampling nested Archimedean copulas

in large dimensions only requires more Exp(1) random variates, which are easy to generate. Finally, let us note that the numbers of precomputed function values of F indicate that F grows slowly, even for comparably small values of Kendall's tau. For a specific application, one should therefore carefully check the quality of the generated data. For example, if one is particularly interested in the upper tail when the parameter ϑ, thus τ, is large, then truncation of F, even at some large point, may lead to inaccuracies in the simulations. This is not surprising since one tries to find realizations of a random variable with expected value equal to infinity. However, truncating the corresponding distribution function forces the drawn variates to be finite and bounded. As mentioned earlier, under such a setup one should take a closer look at other techniques such as asymptotics in the tail or more sophisticated methods. Note that although we are interested in the upper tail of F in Chapter 5, we only encounter small dependencies around $\tau = 0.05$ there. At our largest quantile $x_{\max}$, F is approximately 0.9999995 resulting in truncation of about only one out of two million generated random variates.

4.4.2 Nested Frank and Joe copulas

In this subsection, we sample fully-nested trivariate Frank and Joe copulas of Type (2.5). We compare the performance regarding precision and run time for different parameters, chosen to match Kendall's tau ranging from 0.1 to 0.6. Sampling $F_0 = \mathcal{LS}^{-1}[\psi_0]$ for Frank's family is achieved with the algorithm "LK" of Kemp (1981), sampling $F_{01} = \mathcal{LS}^{-1}[\psi_{01}]$ is done as for F_0 and F_{01} for Joe's family, i.e., by precomputing the function values up to $1 - \varepsilon = 1 - 10^{-5}$ once, at most for 500 000 values, and sampling via truncation.

Table 4.2 shows the results for the fully-nested Frank copulas. The first column lists Kendall's tau for the non-sectorial and sector parts. The second column gives the number $x_{\max}^{\text{in}}$ of precomputed function values until F_{01} reaches our chosen truncation level $1 - \varepsilon = 1 - 10^{-5}$. The next three columns list pairwise sample versions of Kendall's tau, where each row is computed based on a sample of size 100 000. The second last column shows p-values for the Kolmogorov-Smirnov test based on 1 000 samples of size 100 000 as described in Subsection 3.3.1. Only the case $\tau(\tau) = 0.4(0.5)$ leads to rejection of the χ^2 test according to the 5% level. Note that for the combinations $\tau(\tau) = 0.4(0.6)$ and $\tau(\tau) = 0.5(0.6)$, two, respectively four, of the 125 cubes have an expected number of contained variates less than five. Finally, the last column of Table 4.2 contains mean run times $\bar{\kappa}$ taken over all 1 000 samples of size 100 000 to obtain more reliable measurements. The run times indicate that nested Frank copulas can be sampled very fast for the given range of dependencies.

$\tau(\tau)$	$x_{\max}^{\text{in}}$	$\hat\tau_{12}$	$\hat\tau_{13}$	$\hat\tau_{23}$	p-value	$\bar\kappa$
0.1(0.2)	41	0.0982	0.1002	0.1994	0.0586	0.19
0.1(0.3)	126	0.0983	0.1003	0.2989	0.6524	0.19
0.1(0.4)	441	0.0989	0.1008	0.3990	0.3132	0.20
0.1(0.5)	2 105	0.0993	0.1010	0.4994	0.4407	0.20
0.1(0.6)	18 416	0.0997	0.1011	0.5994	0.6699	0.21
0.2(0.3)	98	0.1995	0.2011	0.3019	0.4105	0.22
0.2(0.4)	369	0.1992	0.2008	0.4016	0.3608	0.21
0.2(0.5)	1 818	0.1991	0.2005	0.5011	0.7094	0.22
0.2(0.6)	16 210	0.1989	0.1999	0.6007	0.2212	0.22
0.3(0.4)	280	0.2963	0.2976	0.3973	0.6429	0.23
0.3(0.5)	1 480	0.2968	0.2981	0.4982	0.9988	0.24
0.3(0.6)	13 632	0.2972	0.2985	0.5986	0.4929	0.25
0.4(0.5)	1 050	0.3988	0.4018	0.5000	0.0148	0.28
0.4(0.6)	10 496	0.3993	0.4022	0.6003	0.1379	0.31
0.5(0.6)	6 504	0.4986	0.4980	0.6010	0.2039	0.50

Table 4.2 Precision and mean run times in seconds for fully-nested trivariate Frank copulas based on 100 000 vectors of random variates.

4 Directly sampling nested Archimedean copulas

Table 4.3 contains the results for the fully-nested Joe copulas, obtained with the same settings as for Frank's family. The second and third column refer to the outer and inner distribution functions F_0 and F_{01}, respectively. Note that for each of the cases $\tau(\tau) = 0.3(0.6)$, $\tau(\tau) = 0.4(0.6)$, and $\tau(\tau) = 0.5(0.6)$, two of the 125 cubes have an expected number of contained variates less than five. Concerning the run times for sampling nested Joe copulas, the number of summands to be sampled according to Theorem 4.1.2 (3) becomes large as the dependence increases, and run time increases accordingly. Note that theoretically, this number is infinity for $\vartheta_0 > 1$, but on a computer it is some finite large number. All in all, both the sampling algorithms for nested Frank and nested Joe copulas are fast for a wide range of dependencies. Let us mention again at this point that other methods for sampling the distributions involved may be applied, see Subsection 3.3.1.

$\tau(\tau)$	$x_{\max}^{\text{out}}/F_0$	$x_{\max}^{\text{in}}/F_{01}$	$\hat{\tau}_{12}$	$\hat{\tau}_{13}$	$\hat{\tau}_{23}$	p-value	$\bar{\kappa}$
0.1(0.2)	117 106	145 442	0.0975	0.1035	0.1977	0.4972	0.30
0.1(0.3)	117 106	0.999947	0.0985	0.1040	0.2981	0.7756	0.40
0.1(0.4)	117 106	0.999554	0.0990	0.1039	0.3991	0.2902	0.41
0.1(0.5)	117 106	0.997299	0.0997	0.1039	0.4997	0.9311	0.41
0.1(0.6)	117 106	0.987370	0.1001	0.1036	0.5992	0.3200	0.41
0.2(0.3)	0.999961	191 110	0.2029	0.2053	0.3018	0.8453	0.75
0.2(0.4)	0.999961	0.999923	0.2030	0.2051	0.4015	0.5913	0.85
0.2(0.5)	0.999961	0.999266	0.2027	0.2045	0.5012	0.1045	0.89
0.2(0.6)	0.999961	0.995084	0.2027	0.2042	0.6004	0.7063	0.94
0.3(0.4)	0.999701	273 098	0.3013	0.2984	0.3985	0.2624	2.46
0.3(0.5)	0.999701	0.999876	0.3004	0.2984	0.4989	0.7706	2.67
0.3(0.6)	0.999701	0.998612	0.2999	0.2984	0.5987	0.0804	2.87
0.4(0.5)	0.998330	444 982	0.3982	0.3990	0.5006	0.7215	9.65
0.4(0.6)	0.998330	0.999765	0.3982	0.3990	0.5999	0.9908	10.29
0.5(0.6)	0.992701	0.999984	0.4969	0.4988	0.5983	0.3979	34.77

Table 4.3 Precision and mean run times in seconds for fully-nested trivariate Joe copulas based on 100 000 vectors of random variates.

Figure 4.1 shows two scatter plot matrices of 1 000 vectors of random variates following trivariate fully-nested Frank and Joe copulas, respectively, with parameters chosen to match

4.4 Experiments and examples

$\tau(\tau) = 0.2(0.5)$. Note the radial symmetry for the Frank copula and the upper tail dependence for the Joe copula.

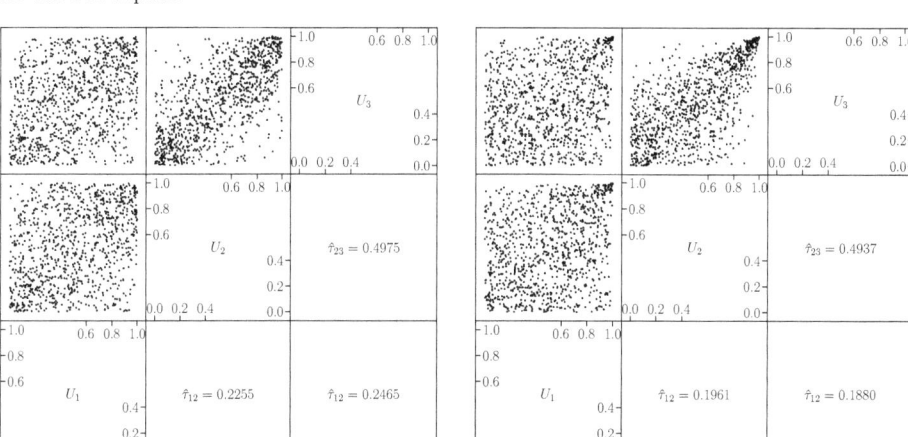

Figure 4.1 1 000 vectors of random variates of a trivariate fully-nested Frank (left) and Joe (right) copula. The parameters involved are chosen such that Kendall's tau of 0.2 and 0.5 are matched.

4.4.3 Comparison of algorithms for nested Clayton copulas

As we have seen in Subsection 3.3.6, sampling nested Clayton copulas is not an easy task. The difficulty lies in the fact that F_{01} is an exponentially tilted stable distribution. In this subsection we investigate the performance of the algorithms of Rosiński (2007), Ridout (2008), the standard rejection algorithm, and our proposed fast rejection algorithm for sampling nested Clayton copulas. Due to the dependence of F_{01} on $V_0 \sim \Gamma(1/\vartheta_0, 1)$, sampling these copulas is especially complicated since we do not have to generate random variates from the same distribution with fixed parameters, but almost surely changing parameters for every vector of random variates to be generated. For this reason, an algorithm which, on the one hand, works fast for sampling a large amount of exponentially tilted stable distributed random variates but which, on the other hand, requires complicated setup steps depending on the parameters of the distribution may not be adequate for sampling nested Clayton copulas. This is essentially what we see in Table 4.4 for 100 000 generated vectors of random variates.

4 Directly sampling nested Archimedean copulas

First consider the procedure of Rosiński (2007). Since V_0 almost surely changes for every random variate V_{01} to be generated, Algorithm 4.2.5 can not be efficiently used. Instead, Representation (4.7) applies. Truncating this series at some $J \in \mathbb{N}$ still requires generating $3J$ random variates for each random variate V_{01} to be drawn. For the computations in Table 4.4 we choose $J = 10\,000$. Although this choice is known to be rather fast than unbiased, see Ridout (2008), the algorithm of Rosiński (2007) turns out to be slow. Note that this can be different if large amounts of exponentially tilted stable distributed random variates with fixed parameters are required. However, we would like to note that much larger J should be used. By considering sample versions of Kendall's tau, one realizes that the choice $J = 10\,000$ is too biased to provide reliable results. For example, if $\tau(\tau) = 0.025(0.6)$ we obtain $\hat{\tau}_{23} = 0.0384$, and $\tau(\tau) = 0.2(0.6)$ results in $\hat{\tau}_{23} = 0.3087$. Similarly for the other cases, where $\hat{\tau}_{23}$ is consistently too small. This clearly indicates that V_{01}, which introduces the sector dependence, is not correctly sampled.

The algorithm of Ridout (2008) requires some setup steps which change with every V_0 obtained from the first step of McNeil's algorithm. This, together with the fact that finding the quantile of a uniform random variate with a modified Newton-Raphson method can be time-consuming, causes the algorithm to be slow. The numbers in parentheses in Table 4.4 are the numbers of warnings we obtained. These warnings indicate that the accuracy for solving $F_{01}(x) = U$ with respect to x for a uniform random variate U could not be reached after the default number of iterations of the modified Newton-Raphson method, which is $1\,000$. Although we had already changed the default precision from 10^{-7} to 10^{-5}, it was not possible in all these cases to obtain x values for which $|F_{01}(x) - U| \le 10^{-5}$ after $1\,000$ iterations. Moreover, for larger dependencies, measured with Kendall's tau, it was not possible to locate a quantile for at least one of the random variates, see the error message "`Cannot locate upper quantile`" in the implementation of this algorithm. These cases are indicated by a bar.

The standard rejection algorithm originally proposed by McNeil (2008) for sampling nested Clayton copulas is an exact algorithm. It requires neither approximations nor numerically complicated setup steps. However, if there is little dependence, the exponential complexity of this algorithm in V_0 forces it to be slow. For some parameter setups it was not possible to generate $100\,000$ vectors of random variates in less than 555 hours, as indicated by a plus sign. Note that after 555 hours, not more than $1\,800$ random variates V_{01} were generated for the cases where the non-sectorial Kendall's tau equals 0.025 and the sectorial one is in $\{0.05, 0.1, 0.2, 0.3, 0.4\}$ and not more than $6\,600$ if the latter one equals 0.5.

Besides the run-time issue, another problem appears. When generating $S(\alpha, 1, (\cos(\alpha\pi/2) \cdot V_0)^{1/\alpha}, V_0 \mathbb{1}_{\{\alpha=1\}}; 1)$ random variates for the case $\tau(\tau) = 0.025(0.6)$, the resulting parameter α

4.4 Experiments and examples

$\tau(\tau)$	Rosiński	Ridout	Standard rejection	Fast rejection
0.025(0.05)	657.45	49.46	+	6.01
0.025(0.1)	658.15	19.92	+	5.78
0.025(0.2)	658.72	14.04	+	5.73
0.025(0.3)	634.66	13.72 (1)	+	5.67
0.025(0.4)	635.39	20.02 (29)	+	5.54
0.025(0.5)	634.71	44.56 (217)	+	5.52
0.025(0.6)	635.11	42.77 (109)	−	5.62*
0.05(0.1)	658.36	20.64	659 649.15	3.06
0.05(0.2)	657.90	14.73	611 434.81	2.95
0.05(0.3)	658.71	14.17 (1)	690 703.44	2.93
0.05(0.4)	634.06	19.13 (30)	697 543.63	2.89
0.05(0.5)	635.37	35.89 (128)	727 002.83	2.88
0.05(0.6)	634.89	45.43 (172)	650 025.72	2.87
0.1(0.2)	658.22	16.82	1 171.66	1.59
0.1(0.3)	658.88	15.22 (1)	1 051.45	1.53
0.1(0.4)	658.79	17.19 (5)	1 288.19	1.53
0.1(0.5)	657.63	27.90 (74)	1 716.79	1.53
0.1(0.6)	634.69	42.11 (148)	1 742.28	1.50
0.2(0.3)	658.60	19.16	15.39	0.87
0.2(0.4)	656.96	17.19 (1)	11.25	0.84
0.2(0.5)	653.88	25.71 (14)	14.85	1.41
0.2(0.6)	658.95	31.59 (79)	23.73	0.82
0.3(0.4)	657.65	22.48 (1)	2.60	0.64
0.3(0.5)	658.91	20.71 (3)	2.77	0.60
0.3(0.6)	658.05	−	2.56	0.60
0.4(0.5)	657.93	−	2.41	0.52
0.4(0.6)	658.80	−	1.85	0.51
0.5(0.6)	658.04	−	1.25	0.90

Table 4.4 Run times in seconds for generating 100 000 vectors of random variates from fully-nested trivariate Clayton copulas with different parameters.

4 Directly sampling nested Archimedean copulas

equals 2/117. Due to this rather small value, some powers which have to be computed in the function `gsl_ran_levy_skew` of the GNU Scientific Library, see GSL, are smaller than the smallest strictly positive floating-point number in `double` precision, see Subsection 3.3.1. Thus, they are treated as zero in computer arithmetic. This leads to zero return values of `gsl_ran_levy_skew`, thus of V_{01}, for the given realizations V_0, for 71 556 out of 100 000 vectors of random variates. We are thus not able to properly generate $\tilde{S}(\alpha, 1, (\cos(\alpha\pi/2)V_0)^{1/\alpha}, V_0\mathbb{1}_{\{\alpha=1\}}, \mathbb{1}_{\{\alpha\neq 1\}}; 1)$ distributed random variates with the standard rejection algorithm in these cases, this is indicated by a bar in Table 4.4. For the fast rejection algorithm, this only occurred for three out of the 100 000 generated vectors of random variates, as indicated by a star. Note that although this is only a computational problem, we feel obliged to address such issues since both researchers and practitioners often do not seem to be aware of them.

The principle underlying the fast rejection algorithm to reduce the complexity from exponential to linear complexity in V_0 significantly decreases run time. As Theorem 4.2.3 implies, for every random variate $V_0 > 1$, the algorithm is uniformly faster than the standard rejection algorithm over all dependencies. Further, the linear complexity in V_0 causes this algorithm to be efficient over the whole range of investigated parameters, and complicated setup steps are also not required. Moreover, this algorithm is exact. Therefore, the fast rejection algorithm is strongly recommended for sampling nested Clayton copulas.

The plot on the left-hand side of Figure 4.3 shows a scatter plot matrix of 1 000 generated vectors of random variates from a fully-nested Clayton copula of Type (2.5) with corresponding pairwise sample versions of Kendall's tau. The parameters are chosen such that $\tau(\tau) = 0.2(0.5)$.

4.4.4 Overall precision and run-time comparison

Table 4.5 contains pairwise sample versions of Kendall's tau and mean run times $\bar{\kappa}_{1(2)}$ for drawing 100 000 vectors of random variates from different fully-nested Archimedean copulas of Type (2.5). Further, mean run times $\bar{\kappa}_{50(50)}$ for the corresponding nested Archimedean copulas with dimensions 50 for each of the non-sectorial and sector parts are given. The mean run times are computed based on 1 000 replications. Our list of nested Archimedean copulas includes the families of Ali-Mikhail-Haq, Clayton, Frank, Gumbel, and Joe, as well as an outer power Clayton copula ("opC") and the family combinations addressed in Theorem 4.3.2. As a benchmark, we include a Gauss copula ("Ga") and a t copula with four degrees of freedom ("t$_4$"), see Algorithm 1.9.5 and Algorithm 1.9.6 concerning sampling strategies. For all examples, the parameters are chosen such that pairwise Kendall's tau are as given in the second column.

4.4 Experiments and examples

The last column of Table 4.5 indicates that all presented algorithms are fast enough to be applied in large-scale simulation studies. This is done in Chapter 5 for partially-nested Archimedean copulas of Type (2.6) with $S = 6$. Further, according to the last two columns of Table 4.5, sampling F_0 and F_{01} does not seem to be the most time-consuming step anymore when the dimension is increased.

Copula	$\tau(\tau)$	$\hat{\tau}_{12}$	$\hat{\tau}_{13}$	$\hat{\tau}_{23}$	$\bar{\kappa}_{1(2)}$	$\bar{\kappa}_{50(50)}$
A	0.2(0.3)	0.1969	0.1997	0.3011	0.15	2.59
C	0.2(0.5)	0.1994	0.1993	0.4965	0.83	5.37
F	0.2(0.5)	0.1991	0.2005	0.5011	0.22	5.15
G	0.2(0.5)	0.1984	0.1965	0.5014	0.46	5.36
J	0.2(0.5)	0.2027	0.2045	0.5012	0.90	6.06
opC	0.2(0.5)	0.1993	0.1974	0.4998	0.55	7.28
A(C)	0.2(0.5)	0.1979	0.1985	0.4976	0.48	5.65
A(19)	0.2(0.5)	0.2032	0.2028	0.4975	0.35	3.33
A(20)	0.2(0.7)	0.2030	0.2051	0.7014	0.91	5.06
C(12)	0.2(0.5)	0.2009	0.2027	0.5034	0.82	6.46
C(14)	0.2(0.5)	0.2016	0.2001	0.5029	1.09	6.71
C(19)	0.2(0.5)	0.2008	0.1987	0.5030	0.81	5.73
C(20)	0.2(0.5)	0.1999	0.1989	0.5000	1.09	6.08
Ga	0.2(0.5)	0.2014	0.1981	0.5000	0.09	7.26
t_4	0.2(0.5)	0.1993	0.1989	0.4993	0.33	14.83

Table 4.5 Precision and mean run times in seconds for drawing 100 000 vectors of random variates from different nested Archimedean copulas.

4.4.5 An Ali-Mikhail-Haq(Clayton(20)) copula

All presented algorithms are also applicable to more than two hierarchies. As an example we sample a four-dimensional A(C(20)) copula, i.e., a fully-nested Archimedean copula based on the families of Ali-Mikhail-Haq, with generator ψ_0, Clayton, with generator ψ_1, and the family numbered 20, with generator ψ_2, on the first, second, and third level, respectively. For the corresponding tree representation, see Figure 4.2.

4 Directly sampling nested Archimedean copulas

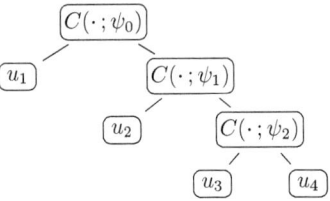

Figure 4.2 Tree structure of a four-dimensional fully-nested Archimedean copula.

For such a four-dimensional fully-nested Archimedean copula, McNeil's algorithm is given as follows, where

$$\psi_{12}(t; V_{01}) := \exp\bigl(-V_{01}\psi_{01}^{-1}(\psi_{02}(t; V_0); V_0)\bigr) = \exp\bigl(-V_{01}\psi_1^{-1}(\psi_2(t))\bigr), \ t \in [0, \infty].$$

Algorithm 4.4.1 (McNeil (2008))

(1) Sample $V_0 \sim F_0 := \mathcal{LS}^{-1}[\psi_0]$.

(2) Sample $X_1 \sim U[0,1]$ independently of V_0.

    (2.1) Sample $V_{01} \sim F_{01} := \mathcal{LS}^{-1}[\psi_{01}(\,\cdot\,; V_0)]$.

    (2.2) Sample $X_2' \sim U[0,1]$ independently of V_0, X_1, and V_{01}.

        (2.2.1) Sample $V_{12} \sim F_{12} := \mathcal{LS}^{-1}[\psi_{12}(\,\cdot\,; V_{01})]$.

        (2.2.2) Sample i.i.d. $X_j'' \sim U[0,1]$, $j \in \{3, 4\}$, independently of V_0, X_1, V_{01}, X_2', and V_{12}.

        (2.2.3) Set $X_j' := \psi_{12}(-\log(X_j'')/V_{12}; V_{01})$, $j \in \{3, 4\}$.

    (2.3) Set $X_j := \psi_{01}(-\log(X_j'); V_0)$, $j \in \{2, 3, 4\}$.

(3) Return $\boldsymbol{U}$, where $U_j := \psi_0(-\log(X_j)/V_0)$, $j \in \{1, \dots, 4\}$.

According to the sufficient nesting condition, the parameters ϑ_i, $i \in \{0, 1, 2\}$, have to satisfy $0 \le \vartheta_0 < 1 \le \vartheta_1 \le \vartheta_2 < \infty$, see Theorem 4.3.2 (1), (4). For sampling the resulting A(C(20)) copula with Algorithm 4.4.1, first note that $V_0 \sim \text{Geo}(1 - \vartheta_0)$. For sampling V_{01}, applying Theorem 4.2.6 leads to the stochastic representation $V_{01} = \tilde{S}V^{\vartheta_1}$ with $V \sim \Gamma(V_0, 1/(1-\vartheta_0))$ and $\tilde{S} \sim \tilde{S}(1/\vartheta_1, 1, \cos^{\vartheta_1}(\pi/(2\vartheta_1)), \mathbb{1}_{\{\vartheta_1=1\}}, V^{\vartheta_1}\mathbb{1}_{\{\vartheta_1\ne 1\}}; 1)$. Sampling V_{12} is addressed in Theorem 4.3.2 (4) and can be achieved with the fast rejection algorithm. For this, note that V_{12} has Laplace-Stieltjes transform $\psi_{12}(t; V_{01}) = \psi(h+t)/\psi(h)$ with $\psi(t) := \exp(-V_0(\log(1+

4.4 Experiments and examples

$t))^\alpha)$, $\alpha := \vartheta_1/\vartheta_2$, $h := e - 1$, and $F_m := \mathcal{LS}^{-1}[\psi^{1/m}] = \Gamma(S_m, 1)$, $m \in \mathbb{N}$, where $S_m \sim$ S$(\alpha, 1, (\cos(\pi\alpha/2)V_{01}/m)^{1/\alpha}, V_{01}\mathbb{1}_{\{\alpha=1\}}/m; 1)$, with m chosen as in Theorem 4.2.3 (2) and $c = e^{V_0}$.

The parameters of the A(C(20)) copula are chosen as $\vartheta_0 = 0.7135$, $\vartheta_1 = 4/3$, and $\vartheta_2 = 1.3773$, matching Kendall's tau of 0.2, 0.4, and 0.7 for the corresponding pairs. As a quality check, we sample 100 000 vectors of random variates from this copula. The resulting pairwise sample versions of Kendall's tau are given by $\hat{\tau}_{12} = 0.2019$, $\hat{\tau}_{13} = 0.1996$, and $\hat{\tau}_{13} = 0.1999$, corresponding to 0.2; $\hat{\tau}_{23} = 0.4001$ and $\hat{\tau}_{24} = 0.3983$, corresponding to 0.4; and $\hat{\tau}_{34} = 0.7014$, corresponding to 0.7.

Let us mention another technical problem we found when sampling V_{12} under this setup, which, again, is due to computer arithmetic. As S_m tends to be small, the gamma distribution for sampling F_m obtains small shape parameters. The problem is that the smaller S_m is, the closer $F_m(x_{\min})$ is to one, where $x_{\min} = 2.2251 \cdot 10^{-308}$ is the smallest strictly positive floating-point number in **double** precision, see Subsection 3.3.1. Although a mathematically correct U-quantile for a $U \sim \mathrm{U}[0, 1]$ with $U < F_m(x_{\min})$ would lie in $(0, x_{\min})$, the gamma random variate algorithm `gsl_ran_gamma`, see GSL, returns zero. Thus, a positive probability of V_{12} being zero occurs. Realizations of the pair $(U_3, U_4)^T$ therefore show a positive probability of being zero in both components. Based on the 100 000 generated vectors of random variates, the probability that $(U_3, U_4)^T = (0, 0)^T$ is estimated as 0.44%. The run time is given as 1.22s. By choosing ϑ_2, thus τ_{34}, larger, this probability also becomes larger. Although being a flaw from a mathematical point of view, it is indeed correct from the computational perspective. This behavior may also be seen from the gamma random variate generator `rgamma` used in the statistical software package R. The gamma random variate generator `g05ffc` from NAG behaves differently. Although it does not return zeros and therefore no flaws are visible from the data, this implementation does not seem to be correct. When generating 100 000 gamma random variates with the shape parameter 0.001 and the procedures `gsl_ran_gamma`, `rgamma`, and `g05ffc`, the former two procedures lead to a sample mean of $9.6887 \cdot 10^{-4}$ and $1.047059 \cdot 10^{-3}$, respectively, being close to the correct mean 0.001. The NAG procedure, however, leads to $2.1610 \cdot 10^{-3}$.

The plot on the right-hand side of Figure 4.3 shows a scatter plot matrix of 1 000 generated vectors of random variates from the specified fully-nested A(C(20)) copula with corresponding pairwise sample versions of Kendall's tau.

4 Directly sampling nested Archimedean copulas

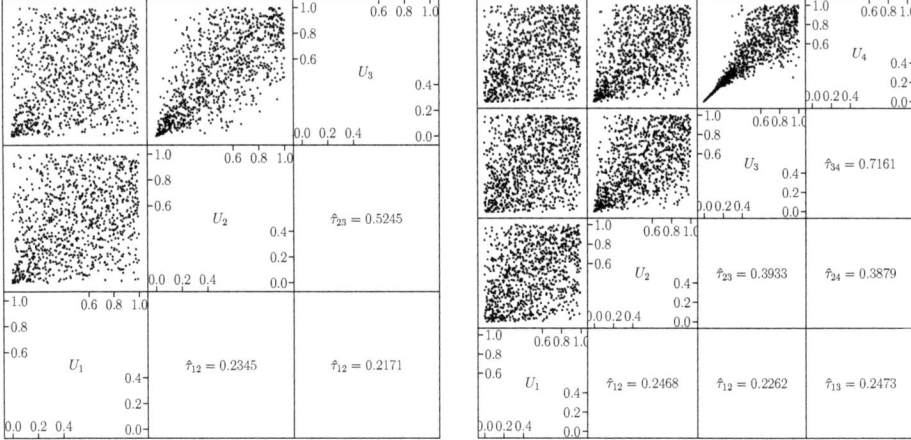

Figure 4.3 1 000 vectors of random variates following a trivariate fully-nested Clayton copula with $\tau(\tau) = 0.2(0.5)$ (left) and a four-dimensional fully-nested A(C(20)) copula with $\tau(\tau(\tau)) = 0.2(0.4(0.7))$ (right).

4.5 Conclusion

McNeil's algorithm is currently the only known algorithm for efficiently sampling nested Archimedean copulas. This elegant algorithm turns out to be particularly challenging as the distribution F_{01} related to the nodes of a nested Archimedean copula depends on the random variate V_0 from the distribution function F_0 corresponding to the outer generator. In particular, V_0 acts as an additional parameter to F_{01}. If F_0 is absolutely continuous, this parameter almost surely changes for every vector of random variates to be generated and possibly quite often if F_0 is discrete. The main idea behind our efficient algorithms is to reduce the influence of V_0 on F_{01}. With this strategy we obtained fast sampling algorithms applicable to virtually all commonly used Archimedean and nested Archimedean copulas. Further, our findings include general strategies for constructing and sampling nested Archimedean copulas.

Our first result addressed the families of Ali-Mikhail-Haq, Frank, and Joe. For these Archimedean families, F_0 is discrete with $V_0 \in \mathbb{N}$. By writing the Laplace-Stieltjes transform ψ_{01} corresponding to F_{01} as an V_0-fold product, one can sample F_{01} as a sum of independent and identically distributed random variates. With this strategy, V_0 enters the scene only in the number of random variates to be drawn. The distribution from which they have to be drawn, however, does not depend on V_0 anymore. Thus, the suggested procedure only requires finding quantiles from a distribution with fixed parameters instead of one which changes for different random variates V_0. This approach can be generalized to any Archimedean family with discrete distribution F_0 on $\mathbb{N}$.

We also introduced a general transformation which allows one to build nested Archimedean copulas based on a given Archimedean generator. For sampling the exponentially tilted stable distributions involved, we developed a fast version of the rejection algorithm. This algorithm applies to tilted Archimedean generators in general. We showed its logarithmic complexity in comparison to the standard rejection algorithm. Nested Archimedean copulas based on several known Archimedean families fall under this setup. Moreover, the presented generator transform and the fast rejection algorithm allow one to start with an Archimedean generator and build, as well as sample, a nested Archimedean copula based on this generator. We compared our proposed algorithm with the ones of Rosiński (2007), Ridout (2008), and the standard rejection algorithm for sampling nested Clayton copulas. In contrast to the former two algorithms, our proposed fast rejection algorithm is exact and does not involve approximations. Further, in contrast to the standard rejection algorithm, it is fast.

By applying the fast rejection algorithm in conjunction with a result on randomized generator powers, we were able to find explicit algorithms for sampling several nested Archimedean copulas based on generators belonging to different Archimedean families. The resulting nested Archimedean copulas even allow for different kinds of pairwise tail dependencies and are therefore quite flexible for modeling purposes.

Our algorithms are promising and the underlying construction and sampling principles may serve as strategies for constructing and sampling other nested Archimedean copulas not addressed in this work or even other distributions given by their Laplace-Stieltjes transforms in a more general context. Moreover, our findings encourage the use of the flexible class of nested Archimedean copulas in large-scale, high-dimensional simulation studies as an alternative to standard elliptical distributions, which are often appreciated for their simple sampling algorithms, but are restricted to be radially symmetric.

4 Directly sampling nested Archimedean copulas

5 CDO pricing with nested Archimedean copulas

> *"Default risk and default correlation can be thought of as consisting of two components: that arising from specific industry or regional problems, and that arising from the general economy."*
>
> Lucas (1995)

The first collateralized debt obligation ("CDO") was issued in 1987. From then on, the global CDO issuance volume rapidly grew from approximately 68 billion USD in 2000 to the astronomical volume of over half a trillion USD in 2006. Then, the global financial crisis came. It turned out that one of the reasons for this crisis was that standard market models failed to adequately price CDOs. This not only seriously hit the CDO market, resulting in the issuance volume to drop to 61 billion USD in 2008 and even less in 2009. It also adversely affected the worldwide financial and economical system.

In this chapter, we present a default model based on nested Archimedean copulas with two nesting levels in order to capture hierarchical dependence structures among the obligors in a credit portfolio. As a main application we treat the pricing of CDOs, for which we show that our approach yields significantly smaller pricing errors for several investigated copulas than the current standard approach using exchangeable copula families.

We start with a brief motivation and introduction to CDOs. Then, we review the intensity-based approach to derive marginal default probabilities. We cover the concept of how dependence is introduced via copulas, emphasize why we apply nested Archimedean copulas in our setup, and explain the implied default correlations. Stylized payment streams of the underlying contracts and the Monte Carlo pricing approach are presented next. To demonstrate the fitting capability of the model, a calibration to CDO tranche spreads of the Markit iTraxx Europe indices is

5 CDO pricing with nested Archimedean copulas

performed. The underlying portfolio consists of credit default swaps on 125 companies from six different industry sectors. It is therefore an excellent portfolio to compare our presented generalized model based on nested Archimedean copulas to a traditional copula model which does not account for different sectors. As such, the standard market model based on the Gaussian copula, as well as several Archimedean copulas, are investigated.

Fast calibration of our model requires fast sampling algorithms for nested Archimedean copulas. Here, our findings of Chapter 4 come into play. Finally, let us note that the content of this chapter was developed in collaboration with Matthias Scherer and accepted for publication, see Hofert and Scherer (2010).

5.1 Motivation and introduction

As observed, e.g., by Lucas (1995), companies in the same industry sector are usually more strongly correlated than firms in different sectors as they are similarly affected by macroeconomic effects, political decisions, or consumer trends. Not considering this valuable information has the unrealistic effect that default correlations do not depend on whether two companies are in the same industry sector or not. A model without sector segmentation is therefore unable to capture empirically observed industry-specific default concentration. Incorporating sector effects to existing credit-portfolio models is straightforward in structural and factor models by introducing sector-specific risk factors. Examples are Hull et al. (2006) or Kiesel and Scherer (2007) for the former, and multi-sector generalizations of Kalemanova et al. (2007) or Albrecher et al. (2007) for the latter. In contrast, tractable portfolio models in the spirit of Schönbucher and Schubert (2001) based on copulas usually utilize specific choices of Archimedean copulas, as the sampling algorithm of Marshall and Olkin (1988) allows one to obtain an approximate loss distribution via a conditionally independent approach. As a drawback, homogeneous pairwise correlations among obligors are inherited from the symmetry of Archimedean copulas.

We introduce an intuitive generalization of the popular copula models of Li (2000) and Schönbucher and Schubert (2001) which allows for asymmetry in form of a hierarchical structure. Our generalization allows us to classify the firms in a credit portfolio according to some attribute, which is the industry sector in our application. Alternative classification criteria, such as geographic regions or political unions, can be used. We achieve this segmentation by using partially-nested Archimedean copulas with two nesting levels. Our model can easily be generalized to hierarchical structures with more than two levels.

As an application of our default model, we treat the pricing of CDOs. In the remaining part

5.1 Motivation and introduction

of this section, we give a brief introduction to these financial products. A good reference to start with is Flanagan and Sam (2002). A more recent reference, including the role CDOs played in the global financial crisis, is Donnelly and Embrechts (2009).

A *collateralized debt obligation* is a type of structured asset-backed security whose payment streams depend on a portfolio of underlying assets exposed to credit risk. CDOs are sophisticated financial products, which are not fully determined until the legal documentation is given. They differ according to the underlying assets (cash CDOs, synthetic CDOs, hybrid CDOs), liabilities (resulting from the tranche structure, coupon payments), purpose of issuance (arbitrage CDOs, balance sheet CDOs), and credit structure (cash flow CDOs, market value CDOs), see Flanagan and Sam (2002).

The main idea of a CDO is to subdivide the credit-risky assets in different risk classes, called *tranches*, which are affected by losses according to their seniority, see Figures 5.1 and 5.2. The process of building these classes is called *securitization*. Securitization makes CDOs attractive for investors with different risk appetites, no matter what the rating of the underlying assets is. Investors seeking speculative-grade assets can invest in the first tranche, called *equity tranche*, and those who seek to put their money into investment-grade assets can invest in more senior tranches. CDO tranches have attracted significant attention among researchers and practitioners, as their spreads, which are explained below, are driven by four different factors: the underlying assets, the individual default probabilities, the joint default probabilities, and the losses in the event of default. Our main concern in this chapter is to adequately model the joint default probabilities, which, according to Duffie (2007), is currently the weakest link in the risk measurement and pricing of CDOs.

In the sequel, we think of a CDO as an insurance against defaults in a portfolio of underlying assets prone to credit risk. The *protection buyer* is obliged to make regular premium payments to the protection seller, whereas the *protection seller* compensates for losses in the portfolio. The premium payments are determined by the CDO tranche spreads and the remaining nominal in the portfolio. The *spread* of a CDO tranche is the annualized fraction which, when multiplied with the amount of money to invest, gives the premium that has to be paid for insuring against defaults affecting the respective tranche. The spread is quoted in *basis points*, i.e., one-hundredth of one percentage point. For example, if we would like to invest 10 000 USD in the second tranche as a protection seller, the spread equals 50 basis points, and the premium payments have to be paid quarterly, we receive a premium of $10\,000\,\text{USD} \cdot 0.005 \cdot 1/4 = 12.5\,\text{USD}$ every three months. This premium is reduced according to the defaulted nominal of the second tranche. For example, after defaults affected the second tranche, such that the remaining nominal equals

5 CDO pricing with nested Archimedean copulas

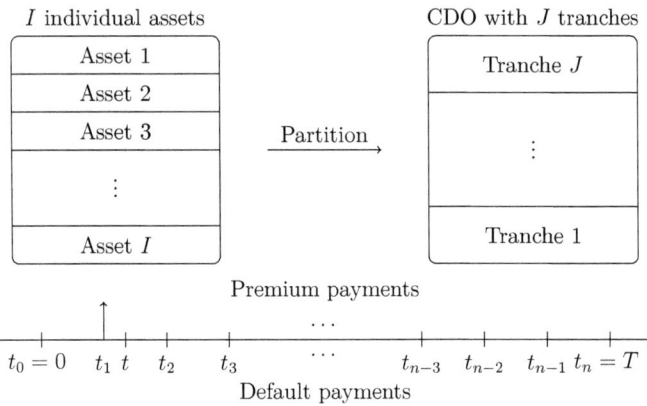

Figure 5.1 A CDO depicted at time t, after the first premium payment was made. So far, no defaults occurred.

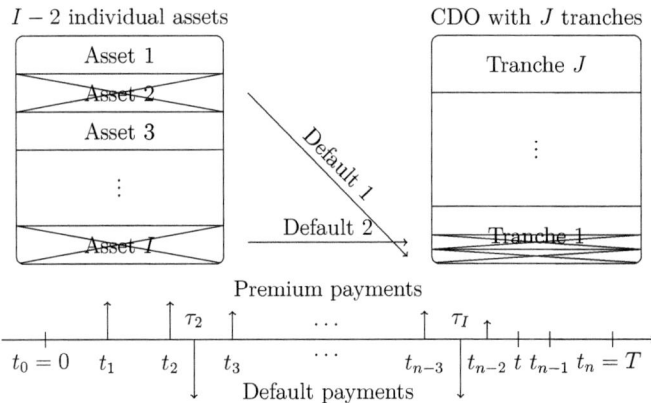

Figure 5.2 A CDO depicted at time t, after two assets are defaulted. Default payments took place and the premium payments are reduced accordingly.

80% of the initial nominal of this tranche, the premium reduces to $12.5\,\text{USD} \cdot 0.8 = 10\,\text{USD}$.

In our applications we treat *synthetic CDOs*, i.e., CDOs which gain exposure to credit-risky assets via derivatives rather than cash purchases of the assets. As derivatives, we consider so-called *credit default swaps* ("CDSs"). A CDS is a credit derivative which allows the (protection) buyer of the CDS to transfer credit risk from of a reference entity to the (protection) seller by agreeing on regular premium payments. The protection seller in turn makes an agreed payoff to the protection buyer in case of a credit event. Note that CDSs allow one to hold the reference entity without being exposed to credit risk. They thus provide a tool to reduce regulatory capital requirements. Finally, let us note that although the first CDS was issued not before 1997, CDS are considered to be among the most important credit derivatives, with a notional value of 45 trillion USD in 2007.

Other financial instruments we encounter in the sequel are *portfolio CDSs*. Up to the precise legal documentation, a portfolio CDS is a synthetic CDO with a single tranche and zero recovery rate.

5.2 The model

The modeled portfolio contains I credit-risky assets whose payment streams depend on the default status of one of I firms. These companies are assigned to one of S industry sectors or classified in one of S sectors according to some attribute. The *default times* of these companies are denoted by τ_i, $i \in \{1, \ldots, I\}$, and are modeled via a simple intensity model. The *intensity* of company i is assumed to be a deterministic, non-negative function, denoted by λ_i. The term structure of the *survival probability* p_i of company i is then given by

$$p_i(t) := \exp\!\left(-\int_0^t \lambda_i(s)\,ds\right), \ t \in [0, \infty], \tag{5.1}$$

and the corresponding *default probability* $\bar{p}_i$ is $\bar{p}_i(t) := 1 - p_i(t)$. Note that other models for describing the individual default probabilities are also possible, as our model is purely designed to explain the dependence among the firms. The canonical construction of τ_i is given by

$$\tau_i := \inf\{t \geq 0 : p_i(t) \leq U_i\}, \tag{5.2}$$

for $U_i \sim \text{U}[0, 1]$, $i \in \{1, \ldots, I\}$, see Bielecki and Rutkowski (2002, pp. 227) or Schönbucher (2003, p. 122). This construction is extremely useful for simulations, since having drawn a random variate U_i, we simply have to compute τ_i via $p_i^{-1}(U_i)$.

5 CDO pricing with nested Archimedean copulas

Indisputably, corporate defaults are not mutually independent. In the presented model, we incorporate the dependence among the default times by making the random variables U_i dependent, in the spirit of Li (2000) and Schönbucher and Schubert (2001). The relevant quantity for pricing portfolio derivatives and risk management purposes is the portfolio-loss process. Given the default time, recovery rate, and nominal of each firm, the portfolio-loss process can be derived. By assuming the vector of *trigger variables* $(U_1, \ldots, U_I)^T$ to be jointly distributed according to a copula C, the resulting distribution of the portfolio-loss process is in general not analytically available. However, for Archimedean copulas it is possible to approximate the portfolio-loss distribution via a conditionally independent approach, see Schönbucher (2003, pp. 340). Unfortunately, this idea can not be generalized to nested Archimedean copulas. Still, by using the algorithms presented in Chapter 4, it is possible to simulate the portfolio-loss process and to price portfolio derivatives by means of a Monte Carlo simulation.

5.2.1 The copulas considered

In the considered framework, company i, $i \in \{1, \ldots, I\}$, defaults up to time t if and only if its trigger variable U_i satisfies $U_i \geq p_i(t)$, i.e., if and only if it is close to one. Therefore, a joint default of companies k and l occurs if their respective trigger variables U_k and U_l are close to one simultaneously. Hence, it is intuitively clear that upper tail dependence plays an important role in our model and that copulas which are able to capture this effect are of interest. Archimedean copulas are able to achieve this. However, their symmetry is usually a drawback as all margins of the same dimension are equal. This, in particular, implies the same correlation among all components. To overcome this shortcoming we suggest to use nested Archimedean copulas.

Due to the lack of closed-form expressions of the portfolio-loss distribution in our hierarchical framework, we have to rely on Monte Carlo techniques for the pricing of portfolio derivatives. The critical step in the simulation of the portfolio-loss process is the fast simulation of the uniformly distributed trigger variables with some described dependence structure. Besides the standard market model using the Gaussian copula, the dependencies we are interested in are partially-nested Archimedean copulas of Type (2.6) with two nesting levels and S sectors of dimensions d_s, $s \in \{1, \ldots, S\}$, where $I = \sum_{s=1}^{S} d_s$ is the dimension, i.e., the number of credit-risky assets in the portfolio. Nested Archimedean copulas of this type, based on the generators ψ_0 on the first level and ψ_s, $s \in \{1, \ldots, S\}$, on the second level, comprise $S+1$ different bivariate margins and are thus much more flexible than the Archimedean copula with generator ψ_0. Assuming that the sufficient nesting condition holds for all nodes involved, i.e., that $(\psi_0^{-1} \circ \psi_s)'$ is completely monotone for all $s \in \{1, \ldots, S\}$, a proper copula results. It can then be sampled with McNeil's algorithm, see Algorithm 2.5.3, for which we apply the strategies developed in Chapter 4. We investigate the Archimedean families of Ali-Mikhail-Haq, Clayton, Frank, Gumbel, and Joe. Due to its flexibility, we also include the outer power Clayton family. A generator for this family is given by $(1+t^{1/\vartheta})^{-1/\vartheta_C}$ with parameter $\vartheta \in [1,\infty)$ and Clayton parameter $\vartheta_C \in (0,\infty)$, assumed to be chosen beforehand. Details about the implied dependence are given in Section 2.4.

5.2.2 Default correlations in a nested framework

As a risk measure for dependent defaults, the concept of default correlation is widely used by risk managers, rating agencies, and regulators. Slightly extending the classical notion of Lucas (1995), the *default correlation* of two companies k and l up to time t is defined as

$$\rho_{kl}(t) := \operatorname{Cor}[\mathbb{1}_{\{\tau_k \leq t\}}, \mathbb{1}_{\{\tau_l \leq t\}}], \ t \in [0, \infty].$$

By assuming identical intensities in Model (5.2), the following result can be derived, see Schönbucher (2003, p. 348). It gives the theoretical term structure of default correlation as well as the corresponding limit as time to maturity decreases to zero.

Theorem 5.2.1
Let C denote the copula of the vector of trigger variables $(U_1, \ldots, U_I)^T$ and let the marginal default probabilities be specified by Construction (5.2) with identical intensities λ and survival probabilities p, i.e., $\lambda_i = \lambda$ and $p_i = p$ for all $i \in \{1, \ldots, I\}$.

(1) The default correlation $\rho_{kl}(t)$ between two companies $k,l \in \{1, \ldots, I\}$ up to time t is given by

$$\rho_{kl}(t) = \frac{C^{(kl)}(p(t), p(t)) - p^2(t)}{p(t)(1-p(t))} = \frac{\hat{C}^{(kl)}(\bar{p}(t), \bar{p}(t)) - \bar{p}^2(t)}{\bar{p}(t)(1-\bar{p}(t))}, \ t \in (0, \infty).$$

(2) For copulas with existing upper tail-dependence coefficient λ_u, $\lim_{t \downarrow 0} \rho_{kl}(t) = \lambda_u$ for all $k,l \in \{1, \ldots, I\}$ with $k \neq l$.

We notice that the original model based on Archimedean copulas implies identical pairwise default correlations for any two firms. In our framework, the pairwise default correlation depends on whether or not two firms are in the same sector. More precisely, we find the following result.

5 CDO pricing with nested Archimedean copulas

Corollary 5.2.2

(1) If C is the partially-nested Archimedean copula of Type (2.6), we have at most $S+1$ different default correlations, given by

$$\rho_s(t) = \frac{\psi_s\bigl(2\psi_s^{-1}(p(t))\bigr) - p^2(t)}{p(t)(1-p(t))}, \quad t \in (0,\infty),$$

depending on whether the two companies under consideration belong to different sectors, take $s=0$, or the same sector, take $s \in \{1,\ldots,S\}$.

(2) For the partially-nested Archimedean copula of Type (2.6), we obtain $\lim_{t\downarrow 0}\rho_s(t) = \lambda_{u,s}$, where the case $s=0$ corresponds to companies belonging to different sectors and $s \in \{1,\ldots,S\}$ corresponds to companies belonging to the same sector. In this formula, $\lambda_{u,s}$ denotes the upper tail-dependence coefficient of the Archimedean copula generated by ψ_s, which is assumed to exist.

Proof

Note that Theorem 5.2.1 only depends on the bivariate marginals of the underlying copula C. For nested Archimedean copulas of Type (2.6), we have at most $S+1$ different bivariate margins, leading to the stated result. $\square$

Corollary 5.2.3

According to Remark 2.3.2, the copulas resulting from the Archimedean families we nest are ordered in the concordance ordering. This implies that larger values of the outer parameter ϑ_0 translate to larger default correlations between companies belonging to different sectors. Similarly, larger values of the inner parameter ϑ_s lead to larger default correlations between companies belonging to the same sector $s \in \{1,\ldots,S\}$. Note that the sufficient nesting condition for any sector s implies that a pair of random variables belonging to the same sector is at least as concordant as a pair of random variables belonging to different sectors. This results in $0 \leq \rho_0(t) \leq \rho_s(t)$ for any $t \in (0,\infty)$ and $s \in \{1,\ldots,S\}$.

5.3 Portfolio CDSs and CDOs

In this section, we present the payment streams of our default model and the algorithms for calibrating it. We also derive two-sided asymptotic confidence intervals for CDO tranche spreads.

5.3.1 The payment streams

Each of the I obligors in the portfolio we consider is assumed to contribute $1/I$ to the unit nominal of the portfolio. The *time to maturity* in years is denoted by T. Premium payments are made according to the *payment schedule* $\mathcal{T} := \{t_0 := 0 < t_1 < \cdots < t_n := T\}$. Defaults may happen anytime in $[0, T]$, but to simplify the computations we defer all default payments which occur between two premium payment dates to the next scheduled payment date. For assessing *accrued interest*, i.e., the interest accumulated between a default and the last payment date, we assume that defaults happen at the midpoint of two payment dates. Therefore, accrued interest for defaulted companies is considered by taking the midpoint $(t_{k-1} + t_k)/2$ as the reference point for a default $\tau_i \in [t_{k-1}, t_k)$. In case of a default, a certain fraction of the value of the reference entity can usually be recovered, which is called *recovery rate*. In what follows we model the payment streams by assuming an identical deterministic recovery rate R for all companies. The *portfolio-loss process* L_t can then be computed by

$$L_t := \frac{1-R}{I} \sum_{i=1}^{I} \mathbb{1}_{\{\tau_i \leq t\}}, \ t \in [0, T]. \tag{5.3}$$

Based on this overall portfolio loss, the *loss affecting the CDO tranche* j, $j \in \{1, \ldots, J\}$, is given by

$$L_{t,j} := \min\{\max\{0, L_t - l_j\}, u_j - l_j\}, \ t \in [0, T], \tag{5.4}$$

where the *lower* and *upper attachment points* l_j and u_j, respectively, mark the endpoints of the CDO tranche j, $j \in \{1, \ldots, J\}$, as a fraction of $[0, 1]$. Given the portfolio loss L_t at $t \in [0, T]$, the *remaining nominal* of the considered portfolio CDS is given by

$$N_t := 1 - \frac{L_t}{1-R}, \ t \in [0, T],$$

i.e., the remaining nominal is reduced by $1/I$ after each default. The *remaining nominal of the CDO tranche* j, $j \in \{1, \ldots, J\}$, is determined by

$$N_{t,j} := u_j - l_j - L_{t,j}, \ t \in [0, T]. \tag{5.5}$$

Let s_T^{pCDS} denote the annualized *portfolio-CDS spread*, quoted in basis points. Further, let $(d_{t_k})_{k=1}^{n}$ denote the *discount factors* corresponding to the time points $(t_k)_{k=1}^{n}$. Given the payment schedule, the portfolio-CDS spread, and the discount factors, the stylized *expected discounted*

5 CDO pricing with nested Archimedean copulas

premium and *default legs* of the portfolio CDS are given by

$$\text{EDPL}_T := \mathbb{E}\left[\sum_{k=1}^n d_{t_k} \Delta_{(t_{k-1},t_k]} s_T^{\text{pCDS}}(N_{t_k} + (N_{t_{k-1}} - N_{t_k})/2)\right], \tag{5.6}$$

$$\text{EDDL}_T := \mathbb{E}\left[\sum_{k=1}^n d_{t_k}(L_{t_k} - L_{t_{k-1}})\right], \tag{5.7}$$

respectively, where $\Delta_{(t_{k-1},t_k]} := t_k - t_{k-1}$ and the summands $d_{t_k}\Delta_{(t_{k-1},t_k]}s_T^{\text{pCDS}}(N_{t_{k-1}} - N_{t_k})/2$ in Definition (5.6) account for accrued interest. The stylized *expected discounted premium* and *default legs of the CDO tranche* j, $j \in \{1,\ldots,J\}$, are given by

$$\text{EDPL}_{T,j} := \mathbb{E}\left[\sum_{k=1}^n d_{t_k}\Delta_{(t_{k-1},t_k]}s_{T,j}^{\text{CDO}}(N_{t_k,j} + (N_{t_{k-1},j} - N_{t_k,j})/2)\right], \tag{5.8}$$

$$\text{EDDL}_{T,j} := \mathbb{E}\left[\sum_{k=1}^n d_{t_k}(L_{t_k,j} - L_{t_{k-1},j})\right], \tag{5.9}$$

respectively, where $s_{T,j}^{\text{CDO}}$ denotes the annualized *CDO tranche spread of tranche* j, $j \in \{1,\ldots,J\}$, quoted in basis points for all but the equity tranche. The *fair spreads* $s_T^{\text{pCDS},f}$ and $s_{T,j}^{\text{CDO},f}$, $j \in \{1,\ldots,J\}$, of the portfolio CDS and the CDO tranches, respectively, are computed by equating the respective expected discounted premium and default legs and solving for the spread. It has become market standard to assume a running spread of 500 basis points for the equity tranche. Therefore, a so-called *upfront payment*, quoted as a percentage of the nominal of the equity tranche, is introduced to correct for this artificial spread. The upfront payment, for simplicity denoted by $s_{T,1}^{\text{CDO}}$ in the sequel, satisfies the relation

$$\text{EDDL}_{T,1} = s_{T,1}^{\text{CDO}}(u_1 - l_1) + \mathbb{E}\left[\sum_{k=1}^n d_{t_k}\Delta_{(t_{k-1},t_k]}0.05(N_{t_k,1} + (N_{t_{k-1},1} - N_{t_k,1})/2)\right]. \tag{5.10}$$

5.3.2 The pricing approach

This section presents the pricing algorithms used to calibrate our model. We begin with the algorithm for pricing portfolio CDSs. If deterministic discount factors are assumed, Definitions (5.6) and (5.7) require only the computation of the expected portfolio loss at the premium-payment dates. Using linearity, this expectation is simply given as the mean of individual default probabilities multiplied by the loss given default, which is easy to compute in our framework. The computation of the expected remaining nominal is done similarly. Thus, the expected discounted premium and default legs can be evaluated for any dependence structure between the companies. This allows one to compute the fair spread of a portfolio CDS via the following algorithm.

5.3 Portfolio CDSs and CDOs

Algorithm 5.3.1 (The fair spread of a portfolio CDS)

(1) *Setup.* Specify the number of companies I, the payment schedule $\mathcal{T}$, the intensity function $\lambda_i(t)$ of each firm, the recovery rate R, and the discount factors d_{t_k}.

(2) *Expected discounted premium and default legs.* Compute the expected discounted premium leg with $s_T^{\text{pCDS}} = 1$ from Definition (5.6) and the default leg from Definition (5.7), respectively, by the argument explained above. Assess the fair spread by

$$s_T^{\text{pCDS},f} = \frac{\text{EDDL}_T}{\text{EDPL}_T}.$$

Principally, the same argument applies to Definitions (5.8) and (5.9) for the different tranches of a CDO. However, as we observe from Definition (5.4), the loss affecting a certain tranche is not a linear functional in the default indicators $\mathbb{1}_{\{\tau_i \leq t\}}$, $i \in \{1, \ldots, I\}$. Now the dependence structure becomes important. Due to the copula combining the trigger variables, it is not straightforward to compute the resulting expected loss of a certain tranche analytically. Therefore, we suggest the following Monte Carlo algorithm for pricing CDO tranches.

Algorithm 5.3.2 (Pricing CDO tranches via Monte Carlo)

(1) *Setup.* Specify I, $\mathcal{T}$, $\lambda_i(t)$, R, and d_{t_k} as in Algorithm 5.3.1. Moreover, specify the number of simulation runs N, the attachment points l_j and u_j of each tranche $j \in \{1, \ldots, J\}$, and a copula C for the trigger variables $(U_1, \ldots, U_I)^T$. If C is the Gaussian copula with identical pairwise correlations, an Archimedean Ali-Mikhail-Haq, Clayton, Frank, Gumbel, Joe, or outer power Clayton copula (with fixed Clayton parameter ϑ_C), only one parameter ϑ has to be chosen. If C is a nested Archimedean copula the parameter vector $\boldsymbol{\vartheta} = (\vartheta_0, \ldots, \vartheta_S)^T$ has to be specified.

(2) *Survival probabilities.* Compute all $p_i(t_k)$, $i \in \{1, \ldots, I\}$, $t_k \in \mathcal{T}$, as in Definition (5.1).

(3) *Monte Carlo simulation.* For each of the N runs, do:

(3.1) Sample the trigger variables $(U_1, \ldots, U_I)^T \sim C$.

(3.2) Compute the corresponding default times via $\tau_i = \min\{t_k \in \mathcal{T} : p_i(t_k) \leq U_i\}$, with the convention that $\min \emptyset := \infty$.

(3.3) Compute the loss L_{t_k}, $t_k \in \mathcal{T}$, of the current Monte Carlo run.

(3.4) For each tranche $j \in \{1, \ldots, J\}$ and each $t_k \in \mathcal{T}$, compute the loss $L_{t_k,j}$ and the remaining nominal $N_{t_k,j}$ via Definitions (5.4) and (5.5), respectively.

5 CDO pricing with nested Archimedean copulas

(4) *Expected discounted premium and default legs.* Based on the Monte Carlo simulation conducted in (3), compute the N discounted premium and default legs for each tranche $j \in \{1, \ldots, J\}$ and estimate the expectations in (5.8) and (5.9) by their sample means $\overline{\mathrm{EDPL}}_{T,j}$ and $\overline{\mathrm{EDDL}}_{T,j}$, respectively. For assessing the fair spreads $s_{T,j}^{\mathrm{CDO},f}$ of all tranches, compute the premium legs for tranche $j \in \{2, \ldots, J\}$ with spread one and estimate $s_{T,j}^{\mathrm{CDO},f}$ by

$$\hat{s}_{T,j}^{\mathrm{CDO},f} := \frac{\overline{\mathrm{EDDL}}_{T,j}}{\overline{\mathrm{EDPL}}_{T,j}}.$$

Finally, determine $\hat{s}_{T,1}^{\mathrm{CDO},f}$ via Equation (5.10).

Estimating CDO tranche spreads via Monte Carlo naturally invokes the question on confidence intervals. Given the straightforward asymptotic confidence intervals for the expected discounted premium and default legs, asymptotic confidence intervals for the tranche spreads are found via the following result.

Proposition 5.3.3 (Two-sided asymptotic confidence intervals for CDO tranche spreads) Given a significance level $\alpha \in (0,1)$ and a CDO tranche $j \in \{1, \ldots, J\}$ with time to maturity T, two-sided asymptotic $(1 - \alpha/2)$-confidence intervals for the expected discounted default and premium legs $\mathrm{EDDL}_{T,j}$ and $\mathrm{EDPL}_{T,j}$ are given by

$$[l_{\mathrm{EDDL}_{T,j}}, u_{\mathrm{EDDL}_{T,j}}] = \left[\overline{\mathrm{EDDL}}_{T,j} - \frac{q_{1-\alpha/4}}{\sqrt{N}} s_{\mathrm{DDL}_{T,j}}, \overline{\mathrm{EDDL}}_{T,j} + \frac{q_{1-\alpha/4}}{\sqrt{N}} s_{\mathrm{DDL}_{T,j}}\right],$$

$$[l_{\mathrm{EDPL}_{T,j}}, u_{\mathrm{EDPL}_{T,j}}] = \left[\overline{\mathrm{EDPL}}_{T,j} - \frac{q_{1-\alpha/4}}{\sqrt{N}} s_{\mathrm{DPL}_{T,j}}, \overline{\mathrm{EDPL}}_{T,j} + \frac{q_{1-\alpha/4}}{\sqrt{N}} s_{\mathrm{DPL}_{T,j}}\right],$$

respectively, where $s_{\mathrm{DDL}_{T,j}}$ and $s_{\mathrm{DPL}_{T,j}}$ denote the sample standard deviations for the simulated discounted default and premium legs and $q_{1-\alpha/4}$ denotes the $(1-\alpha/4)$-quantile of the standard normal distribution. This implies that

$$\left[\frac{l_{\mathrm{EDDL}_{T,j}}}{u_{\mathrm{EDPL}_{T,j}}}, \frac{u_{\mathrm{EDDL}_{T,j}}}{l_{\mathrm{EDPL}_{T,j}}}\right]$$

is a two-sided asymptotic $(1-\alpha)$-confidence interval for the fair spread $s_{T,j}^{\mathrm{CDO},f}$ of tranche j, $j \in \{2, \ldots, J\}$. For the upfront payment, a two-sided asymptotic $(1-\alpha)$-confidence interval is given by

$$\left[\bar{Y} - \frac{q_{1-\alpha/2}}{\sqrt{N}} s_Y, \bar{Y} + \frac{q_{1-\alpha/2}}{\sqrt{N}} s_Y\right],$$

where

$$Y_l := \frac{1}{u_1 - l_1} \sum_{k=1}^{n} d_{t_k}(L_{t_k,1} - L_{t_{k-1},1} - \Delta_{(t_{k-1},t_k]} 0.05(N_{t_k,1} + (N_{t_{k-1},1} - N_{t_k,1})/2)),$$

$l \in \{1, \ldots, N\}$, and $\bar{Y}$ and s_Y denote the sample mean and standard deviation of the variates Y_l, respectively.

Proof

The proofs of the claims about the asymptotic confidence intervals for $\text{EDDL}_{T,j}$ and $\text{EDPL}_{T,j}$, as well as for the upfront payment, involve straightforward applications of Slutsky's Theorem and the Central Limit Theorem. For the statement about $[l_{\text{EDDL}_{T,j}}/u_{\text{EDPL}_{T,j}}, u_{\text{EDDL}_{T,j}}/l_{\text{EDPL}_{T,j}}]$, apply the lower Fréchet bound W and use the results about the asymptotic confidence intervals for $\text{EDDL}_{T,j}$ and $\text{EDPL}_{T,j}$. □

5.4 Calibration to market spreads

One difficulty in the calibration of a portfolio model to CDO quotes is that the upfront payment for the equity tranche complicates the comparison of pricing errors in this tranche to pricing errors in more senior tranches. Combining upfront and running spread, by far the greatest spread is paid for the equity tranche. We therefore follow Kalemanova et al. (2007), Albrecher et al. (2007), and Kiesel and Scherer (2007) and calibrate our model to match the upfront payment (up to bid-ask spreads reflected by ε_{CDO}) of the equity tranche and minimize the distance of the *CDO market spreads* $s_{T,j}^{\text{CDO},m}$, $j \in \{1, \ldots, J\}$, to the model spreads over the remaining tranches. Thus, we aim for

$$D_1 := \left| s_{T,1}^{\text{CDO},f} - s_{T,1}^{\text{CDO},m} \right| \leq \varepsilon_{\text{CDO}} \text{ and } D_2 := \sum_{j=2}^{J} \left| s_{T,j}^{\text{CDO},f} - s_{T,j}^{\text{CDO},m} \right| \to \min, \quad (5.11)$$

where the minimization is taken over the copula parameters involved.

A major advantage of our framework is that the individual default probabilities and the dependence structure are specified independently of each other. This allows us to calibrate our model in two steps, initially fitting the default probabilities to portfolio CDS market quotes by adjusting the intensities and then fitting the dependence structure to CDO market quotes according to (5.11), by appropriately setting the parameters of the copula under consideration. The first step can be achieved with Algorithm 5.4.1, the second with Algorithm 5.4.2, both are given in Subsection 5.4.2 below.

5 CDO pricing with nested Archimedean copulas

5.4.1 Data and setup

We calibrate our model to portfolio CDS and CDO market quotes of the seventh series of the Markit iTraxx Europe indices. The *portfolio CDS market spread* is denoted by $s_T^{\text{pCDS},m}$. Portfolio CDS spreads are available for contracts maturing in three, five, seven, and ten years. Spreads of the first five CDO tranches are traded for the maturities five, seven, and ten years. In both cases, the most liquidly traded ones are the five and ten year contracts. We calibrated our model to the data as at 2007-06-12, 2007-06-14, 2007-06-19, 2007-06-21, and 2007-06-26. All data was retrieved from the Bloomberg database. The Markit iTraxx Europe indices imply a quarterly payment schedule $\mathcal{T}$, i.e., $n = 4T$, where we consider $T \in \{5, 10\}$. The underlying portfolio consists of CDSs on $I = 125$ companies which are mapped to one of the following six business sectors: Auto (10), Consumer (30), Energy (20), Financials (25), Industrials (20), and TMT (20). We refer the interested reader to the website of Markit, see Markit, for the official terms of contract of all derivatives involved, the names of the individual companies, and further information. We assume a homogeneous portfolio with piecewise constant intensities of the form

$$\lambda(t) := \lambda_i(t) := \lambda_5 \mathbb{1}_{[0,5]}(t) + \lambda_{10} \mathbb{1}_{(5,10]}(t), \ t \in [0, 10],$$

for each $i \in \{1, \ldots, I\}$, where in the sequel λ_5 and λ_{10} denote the constant intensities on the intervals $[0, 5]$ and $(5, 10]$, respectively, chosen to match the most liquidly traded contracts. The recovery rate is chosen as $R = 40\%$ for all firms, a commonly accepted assumption. Note that more general assumptions on the marginals might further improve all calibration results. The attachment points of the five traded tranches of the Markit iTraxx Europe indices are given by [0%,3%], [3%,6%], [6%,9%], [9%,12%], and [12%,22%]. The continuously compounded interest rates are derived from par yields by the standard bootstrap method, see Hull (2008, pp. 80). Interest rates corresponding to non-integer maturities are linearly interpolated.

5.4.2 Calibration of the model

As mentioned above, the first step in the calibration of our model consists of fitting the intensities to portfolio CDS market quotes, which can be done with the following algorithm based on Algorithm 5.3.1.

Algorithm 5.4.1 (Fitting intensities via Algorithm 5.3.1)

(1) *Setup.* Specify the parameters as described in Subsection 5.4.1 according to the Markit iTraxx Europe conventions.

(2) *Fitting λ_5 and λ_{10}.* Use a numerical root-finding procedure to find the estimates $\hat{\lambda}_5$ such that $s_5^{\mathrm{pCDS},f} = s_5^{\mathrm{pCDS},m}$. Note that the model spread $s_5^{\mathrm{pCDS},f}$ is increasing in λ_5 and can be computed using Algorithm 5.3.1 with $T = 5$. Given $\hat{\lambda}_5$, use the numerical root-finding procedure a second time to find the estimate $\hat{\lambda}_{10}$ satisfying $s_{10}^{\mathrm{pCDS},f} = s_{10}^{\mathrm{pCDS},m}$. For this, the model spread $s_{10}^{\mathrm{pCDS},f}$ is computed as a function of λ_{10} with fixed $\lambda_5 = \hat{\lambda}_5$ using Algorithm 5.3.1 with $T = 10$.

The dependence structure of our model is specified by the copula from which the trigger variables are drawn. We use the Archimedean families of Ali-Mikhail-Haq, Clayton, Frank, Gumbel, Joe, and an outer power Clayton copula with fixed Clayton parameter $\vartheta_C = 0.1$, each in their exchangeable and nested version. The reason for this choice of ϑ_C is that we prefer a rather small value in order to be able to capture a large interval of possible upper tail-dependence coefficients for the survival Clayton copula. Apart from that, it is arbitrary. For each except for Frank's family we also consider the corresponding survival copulas, see Remark 1.5.2 for a stochastic representation and therefore a sampling recipe. Note that Frank's Archimedean copula family is radially symmetric. As a benchmark for our studies, we include the market standard Gaussian copula, see Algorithm 1.9.5 for sampling purposes. Besides the exchangeable, i.e., equicorrelated, case, we use a sectorial correlation matrix parametrized such that firms in the same sector $s \in \{1, \ldots, S\}$ have correlation ϑ_s and firms in different sectors have correlation ϑ_0. Concerning the implementation, the guidelines given in Subsection 3.3.1 apply. The distribution functions involved in sampling the Frank and Joe copulas are precomputed up to the truncation level $1 - \varepsilon = 1 - 10^{-8}$, but at most for the first $500\,000$ positive integers.

For each of these copula families, our goal is two-fold. We first test if the exchangeable copula is able to produce sufficient dependence to match the quoted upfront payment of the equity tranche. If not, we conclude that the respective copula family is not suitable for modeling CDOs in our framework. If so, we try to improve the fitting quality in a second step by using the nested copula of the same family. To decrease the dimension of the parameter space in the second step, we assume identical parameters $\vartheta_s =: \vartheta_1$ for all sectors $s \in \{1, \ldots, S\}$, resulting in a two-parameter model for the dependence structure. The general model with $S+1$ parameters might capture market quotes even better. However, the calibration requires much more computational time.

The objective function for assessing the fitting quality of our model is stated in (5.11). Since fair spreads are found by simulation we develop our own optimizer. This routine exploits the specific structure of the problem to achieve the required precision over a relatively short timeframe. The idea of our routine is to first choose the parameter ϑ of the exchangeable copula

5 CDO pricing with nested Archimedean copulas

such that the quoted upfront payment is matched. This position is denoted by $\hat{\vartheta}$. The second step starts with position $(\hat{\vartheta}, \hat{\vartheta})^T$ with the nested copula of the same class and follows the level curve satisfying $D_1 \leq \varepsilon_{\text{CDO}}$ on a fine two-dimensional grid. Here we use the fact that this level curve is decreasing in the $\vartheta_0 \vartheta_1$-plane with $\vartheta_0 \leq \vartheta_1$. To understand this, note that the expected overall loss affecting the portfolio remains the same no matter what the particular dependence structure imposed by the trigger variables is, see Definition (5.3). However, the loss affecting a certain tranche is influenced by the dependence structure. If the joint copula of two trigger variables is one of our chosen copulas, then a larger parameter ϑ implies more positive lower orthant dependence, see Remark 2.3.2 and Joe (1997, p. 50). By Theorem 5.2.1, default correlation is increasing in positive lower orthant dependence of the copula under consideration. A larger default correlation in turn implies a smaller upfront payment, thus the level curve on which the model matches the quoted upfront payment is decreasing. This justifies our CDO calibration algorithm. To see that the upfront payment is decreasing in default correlation, as well as the spread of the most senior tranche is increasing in default correlation, note that the larger the default correlation, the more the portfolio behaves like a single credit, i.e., the probabilities that the equity tranche and the most senior tranche are affected by losses approach each other.

Algorithm 5.4.2 (CDO calibration)

(1) *Setup.* In our calibration we use parameters as described in the beginning of Section 5.4 and Subsection 5.4.1. We further specify a copula C for the trigger variables $(U_1, \ldots, U_I)^T$, which we assume to be one of the copulas mentioned above. Denote by $[\vartheta_l, \vartheta_u]$ the parameter space for the optimization of ϑ, depending on the chosen family. Choose the number of sectors S (e.g., 6) and companies in each sector (e.g., $d_1 = 10$, $d_2 = 30$, $d_3 = 20$, $d_4 = 25$, $d_5 = 20$, and $d_6 = 20$), in correspondence to the Markit iTraxx Europe specifications. Specify ε_{CDO} (e.g., 0.0004) as the accepted pricing error for the upfront payment of the first tranche and choose the number N (e.g., 500 000) of simulation runs. Specify the number m_1 of subdivisions of each dimension for the 2d-optimizer (e.g., 200, 300, or 400, depending on the length $\vartheta_u - \vartheta_l$ of the starting interval) and the number m_2 (e.g., 3) of subdivisions for the refinement step.

(2) *Fitting λ_5 and λ_{10}.* Calibrate the model to match portfolio CDS spreads. For this, use Algorithm 5.4.1 to obtain the fitted intensities $\hat{\lambda}_5$ and $\hat{\lambda}_{10}$.

(3) *1d-optimization.* For the parameter $\vartheta \in [\vartheta_l, \vartheta_u]$ of the one-parameter copula C find a value $\hat{\vartheta}$ satisfying $D_1 \leq \varepsilon_{\text{CDO}}$, see (5.11), by using a bisection procedure. If there is no such parameter, stop and conclude that C is not adequate for our modeling purpose.

(4) *2d-optimization.* If C is a nested copula use $\boldsymbol{\vartheta}_0 := (\vartheta_0, \vartheta_1)^T := (\hat{\vartheta}, \hat{\vartheta})^T$ from the 1d-optimization in (3) as initial vector for the minimization of D_2 over all $(\vartheta_0, \vartheta_1)^T \in [\vartheta_l, \vartheta_u]^2$ satisfying the constraint $D_1 \leq \varepsilon_{\text{CDO}}$. Note that being the result of the 1d-optimization, the vector $(\hat{\vartheta}, \hat{\vartheta})^T$ implies $D_1 \leq \varepsilon_{\text{CDO}}$. For the minimization of D_2, define a fine grid on the parameter space $[\vartheta_l, \vartheta_u]^2$ with mesh $l := (\vartheta_u - \vartheta_l)/m_1$.

(4.1) For each of the parameter constellations $\boldsymbol{\vartheta}_1 := (\vartheta_0 - l, \vartheta_1)^T$, $\boldsymbol{\vartheta}_2 := (\vartheta_0 - l, \vartheta_1 + l)^T$, and $\boldsymbol{\vartheta}_3 := (\vartheta_0, \vartheta_1 + l)^T$ derive CDO tranche spreads with Algorithm 5.3.2.

(4.2) If the upfront payment of at least one of the parameter vectors $\boldsymbol{\vartheta}_k$, $k \in \{1, 2, 3\}$, is found to satisfy $D_1 \leq \varepsilon_{\text{CDO}}$, set $\boldsymbol{\vartheta}_0$ to the vector that minimizes D_1. Otherwise, consider the direction $\boldsymbol{\vartheta}_k$ minimizing D_1 and subdivide the segment from $\boldsymbol{\vartheta}_0$ to $\boldsymbol{\vartheta}_k$ into m_2 equally spaced parts. For all except the previously computed endpoints of this segment, apply Algorithm 5.3.2 to derive CDO tranche spreads for the corresponding $m_2 - 1$ midpoints, denoted by $\boldsymbol{\vartheta}_l$, $l \in \{4, \ldots, m_2 + 2\}$. Set $\boldsymbol{\vartheta}_0$ to $\boldsymbol{\vartheta}_l$, $l \in \{k, 4, \ldots, m_2 + 2\}$, such that D_1 is minimized.

Repeat (4.1) and (4.2) until $\boldsymbol{\vartheta}_0 \notin [\vartheta_l, \vartheta_u]^2$, i.e., $\vartheta_0 \leq \vartheta_l$ or $\vartheta_1 \geq \vartheta_u$. Given all visited vectors $(\vartheta_0, \vartheta_1)^T$ satisfying $D_1 \leq \varepsilon_{\text{CDO}}$, choose $\hat{\boldsymbol{\vartheta}} := (\hat{\vartheta}_0, \hat{\vartheta}_1)^T$ to be the minimizer of D_2.

5.4.3 Results of the calibration

Tables 5.1 and 5.2 list the calibration results as at 2007-06-12 for the maturities $T = 5$ and $T = 10$, respectively. The results for the other trading days considered are similar and can be found in Chapter B. The average results for all trading days are presented in condensed form in Table 5.3. In all tables, we identify the Archimedean families investigated by their respective symbols. A hat is used to denote the corresponding survival copula. The results in the upper and lower row of the given family denote the exchangeable and nested case, respectively.

Considering the calibration results, the Gaussian copula was outperformed by several Archimedean copulas. In general, families which are able to capture upper tail dependence provided good calibration results for all analyzed trading days and maturities. Our generalization to nested Archimedean copulas reduced the pricing errors for all trading days and maturities. For families which are upper tail dependent the error was reduced significantly. We emphasize that this improvement is already obtained by introducing a single additional parameter. Overall, the outer power Clayton copula provided the most accurate fit to market quotes. For this copula, Table 5.4 lists 98% confidence intervals for the upfront payment and fair spreads computed

5 CDO pricing with nested Archimedean copulas

based on 500 000 runs. The upper and lower row again denote the exchangeable and nested case, respectively. As we may infer from the run times listed in Table 5.3, the concern that the generalization to nested Archimedean copulas is computationally too expensive is not justified. Here, $\bar{\kappa}$ denotes the mean run time in seconds for the respective optimal parameters, taken over all five trading days.

Figure 5.3 shows the implied default correlations for all fitted copulas for the first day considered and time to maturity $T = 5$. We notice the large variety of implied term structures of default correlations. Considering the calibration results allows us to assign the families to three classes. The class that performed the best consists of the outer power Clayton copula, the families of Gumbel, and Joe. Each of these families is able to capture upper tail dependence, a fact which is reflected in the default correlations starting above zero. The implied default correlations of these families are relatively constant over time, which is desirable if CDOs with nonstandard maturities, e.g., four years, have to be priced. Also, the absolute level of implied default correlation, the difference of intra- to inter-sector correlations, and the improvement in fitting quality of the nested compared to the exchangeable Archimedean copula family is similar for the members of this class. The second class encompasses Ali-Mikhail-Haq's family, Clayton's family, its corresponding survival copula family, the family of Frank, the survival copula family based on Gumbel's family, and the Gaussian copula. Except for the survival Clayton copula, these copula families are not able to capture upper tail dependence, which implies vanishing default correlations at time zero and forces the term structure of default correlations to increase over time. Although the survival Clayton copula is theoretically able to capture upper tail dependence, the fitted parameters imply only negligible upper tail dependence. We may infer from Tables 5.1, 5.2, and 5.3 that the nested copulas of this second class perform only slightly better than their exchangeable counterparts. Also, the difference of intra- to inter-sector correlations is relatively small. The last class of copulas consists only of the survival Ali-Mikhail-Haq copula. Like most of the members of the second class, this copula does not allow for upper tail dependence. However, the improvement in fitting quality, as well as the difference of intra- to inter-sector correlations, is significant.

Figure 5.4 illustrates the effectiveness of our CDO calibration algorithm using the outer power Clayton copula as an example. The relevant parameter space consists of all vectors $(\vartheta_0, \vartheta_1)^T$ satisfying $\vartheta_i \in [1, \infty)$, $i \in \{0, 1\}$, and $\vartheta_0 \leq \vartheta_1$.

5.4 Calibration to market spreads

2007-06-12	Dependence		CDO upfront and spreads $\hat{s}_{5,j}^{\text{CDO},f}$					Error
Copula	$\hat{\vartheta}$	ρ in %	$j=1$	$j=2$	$j=3$	$j=4$	$j=5$	D_2
A	– – –	– – –	– –	– –	– –	– –	– –	– –
$\hat{\text{A}}$	0.69 0.12 0.94	3.55 0.23 17.88	7.42 7.42	123.06 108.12	2.13 11.38	0.00 0.82	0.00 0.02	94.95 69.93
C	1.94 1.87 2.03	3.32 3.21 3.48	7.40 7.42	98.67 97.02	19.27 18.21	3.84 3.50	0.26 0.24	64.21 61.84
$\hat{\text{C}}$	0.08 0.08 0.09	3.28 3.15 3.42	7.40 7.42	91.79 88.80	21.41 19.89	5.76 5.22	0.64 0.63	58.32 53.28
opC	1.06 1.05 1.08	8.07 6.10 10.44	7.42 7.43	37.31 44.88	17.51 19.77	11.73 11.42	6.96 5.37	24.86 17.64
$\hat{\text{opC}}$	– – –	– – –	– –	– –	– –	– –	– –	– –
F	2.77 2.72 2.83	3.28 3.22 3.36	7.44 7.42	92.97 90.74	22.34 20.92	5.98 5.26	0.64 0.55	60.65 56.38
G	1.06 1.05 1.09	8.19 5.93 11.10	7.42 7.38	35.94 45.32	17.58 20.66	11.62 11.83	6.81 5.33	26.04 18.86
$\hat{\text{G}}$	1.24 1.24 1.27	3.24 3.09 3.73	7.43 7.39	109.41 105.17	13.10 12.45	0.90 0.89	0.01 0.01	71.96 67.08
J	1.07 1.04 1.10	8.45 5.41 11.83	7.37 7.44	32.01 45.38	17.06 21.24	11.88 11.80	7.24 5.16	30.14 19.31
$\hat{\text{J}}$	– – –	– – –	– –	– –	– –	– –	– –	– –
Ga	0.20 0.16 0.33	3.31 2.49 6.99	7.43 7.37	89.85 89.44	21.68 20.51	5.94 5.52	0.76 0.70	56.71 54.77
Market			7.40	45.12	11.84	5.18	2.14	

Table 5.1 Results as at 2007-06-12 based on 500 000 runs and $T=5$.

5 CDO pricing with nested Archimedean copulas

2007-06-12	Dependence		CDO upfront and spreads $\hat{s}_{10,j}^{\mathrm{CDO},f}$					Error
Copula	$\hat{\vartheta}$	ρ in %	$j=1$	$j=2$	$j=3$	$j=4$	$j=5$	D_2
A	0.71	4.79	36.86	367.06	150.76	62.84	11.77	129.48
	0.70 0.73	4.72 4.92	36.87	365.35	149.86	62.57	11.58	126.81
$\hat{\mathrm{A}}$	0.27	2.41	36.88	536.29	108.22	4.04	0.01	286.13
	0.01 0.83	0.06 19.59	36.91	427.53	150.32	41.73	2.99	178.79
C	0.62	4.30	36.91	388.18	150.82	55.82	8.08	147.34
	0.62 0.62	4.29 4.30	36.85	386.60	150.09	55.77	8.19	144.88
$\hat{\mathrm{C}}$	0.07	4.07	36.87	390.46	146.77	54.33	8.59	143.58
	0.07 0.09	3.87 4.84	36.92	389.40	145.19	53.45	8.47	140.16
opC	1.10	12.58	36.86	279.29	89.68	47.39	24.35	57.10
	1.05 1.16	6.68 17.92	36.92	315.12	115.26	57.25	19.13	44.13
o$\hat{\mathrm{p}}$C	−	−	−	−	−	−	−	−
	− −	− −	−	−	−	−	−	−
F	1.56	6.06	36.91	327.08	143.55	70.88	20.08	95.40
	1.55 1.56	6.02 6.10	36.85	325.38	141.58	69.84	19.66	90.28
G	1.12	14.28	36.90	248.17	85.72	50.78	27.66	98.88
	1.05 1.20	6.65 21.06	36.92	296.78	115.70	61.22	20.70	68.46
$\hat{\mathrm{G}}$	1.10	2.97	36.89	466.13	146.36	22.38	0.40	235.38
	1.09 1.19	2.58 5.67	36.92	451.47	147.39	29.54	1.03	213.95
J	1.17	16.56	36.89	213.34	76.36	50.83	30.52	145.99
	1.02 1.29	2.66 25.72	36.89	313.24	133.75	70.58	17.17	75.86
$\hat{\mathrm{J}}$	−	−	−	−	−	−	−	−
	− −	− −	−	−	−	−	−	−
Ga	0.13	4.36	36.89	381.96	146.96	57.13	10.02	136.62
	0.13 0.16	4.15 5.43	36.87	381.17	146.16	56.53	10.01	134.45
Market			36.88	316.90	93.36	42.53	13.39	

Table 5.2 Results as at 2007-06-12 based on 500 000 runs and $T = 10$.

5.4 Calibration to market spreads

Copula	$T=5$ $\bar{\kappa}$	$\bar{\rho}$ in %	$\tilde{D}_2$	$T=10$ $\bar{\kappa}$	$\bar{\rho}$ in %	$\tilde{D}_2$
A	–	–	–	28.76	4.94	138.14
	–	–	–	47.42	4.78 5.56	134.97
Â	18.84	3.44	91.82	29.34	2.46	303.47
	29.53	0.25 16.96	67.53	37.92	0.08 20.23	189.67
C	30.19	3.18	59.35	40.81	4.50	158.12
	59.93	3.04 3.33	56.06	80.94	4.40 4.65	154.57
Ĉ	30.13	3.12	52.51	40.55	4.20	153.31
	200.13	2.99 3.32	48.95	220.29	4.12 4.40	149.50
opC	45.01	7.73	27.06	55.13	12.90	57.28
	70.21	4.88 11.24	16.18	80.16	7.77 17.53	47.98
F	23.86	3.14	54.71	34.23	6.30	101.86
	61.59	3.04 3.26	50.54	69.53	6.22 6.38	97.84
G	34.14	7.83	29.18	44.34	14.63	99.35
	58.00	4.52 11.93	17.44	68.61	7.53 20.88	72.70
Ĝ	34.65	3.10	69.72	44.86	3.05	246.46
	58.41	2.88 3.93	66.10	69.11	2.67 5.68	223.42
J	33.68	8.09	32.54	43.91	16.98	147.90
	55.50	4.22 12.51	18.11	65.99	3.33 25.84	80.84
Ga	63.74	3.15	51.24	73.83	4.53	145.65
	63.73	2.87 4.21	48.92	73.90	4.29 5.67	143.36

Table 5.3 Average calibration results based on 500 000 runs.

5 CDO pricing with nested Archimedean copulas

2007-06-12	\multicolumn{5}{c}{Two-sided 98% confidence intervals for $s_{T,j}^{\mathrm{CDO},f}$ for opC}				
T	$j=1$	$j=2$	$j=3$	$j=4$	$j=5$
5	[7.33,7.51]	[36.40,38.21]	[16.85,18.17]	[11.18,12.28]	[6.55,7.37]
5	[7.34,7.52]	[43.88,45.87]	[19.07,20.46]	[10.89,11.95]	[5.02,5.72]
10	[36.76,36.95]	[277.79,280.80]	[88.72,90.65]	[46.65,48.12]	[23.83,24.87]
10	[36.82,37.02]	[313.47,316.77]	[114.16,116.35]	[56.46,58.04]	[18.71,19.56]

Table 5.4 Confidence intervals for the CDO upfront payment and spreads for the outer power Clayton copula fitted to the data as at 2007-06-12 based on 500 000 runs.

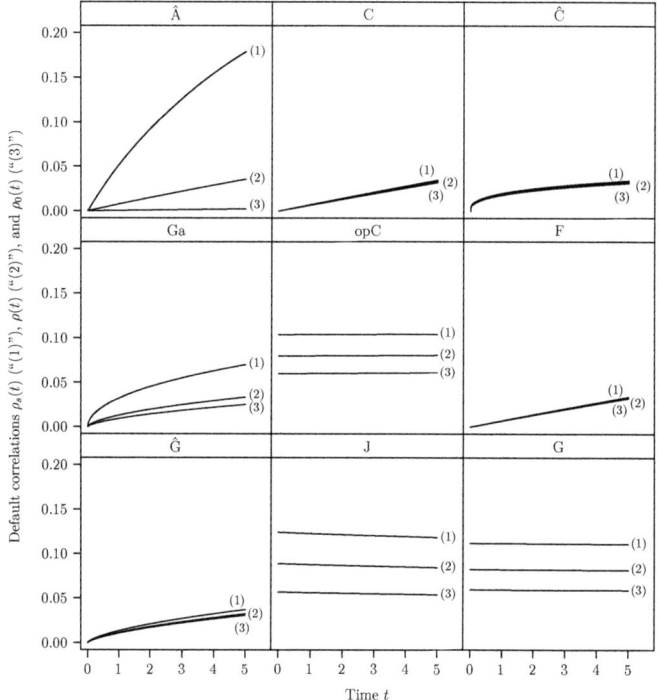

Figure 5.3 Default correlations for the fitted copulas and $T=5$ as at 2007-06-12.

5.4 Calibration to market spreads

At first, a bisection procedure is used to find the parameter $\hat{\vartheta}$ of the exchangeable copula for which the quoted upfront payment is matched. This position is marked with a square. The points considered in this bisection procedure are interpreted as points on the diagonal $\vartheta_0 = \vartheta_1$ within the parameter space. Then, starting from this optimal point on the diagonal, the two-dimensional optimizer follows the level curve on which the nested copula matches the upfront payment. For all visited points on the level curve we compute the errors D_1 and D_2. The former is illustrated by crosses and circles, depending on whether the upfront criterion $D_1 \leq \varepsilon_{\text{CDO}}$ is met or not, respectively, and the latter by different shades of gray. The minimizing argument for the nested Archimedean copula is indicated by a diamond.

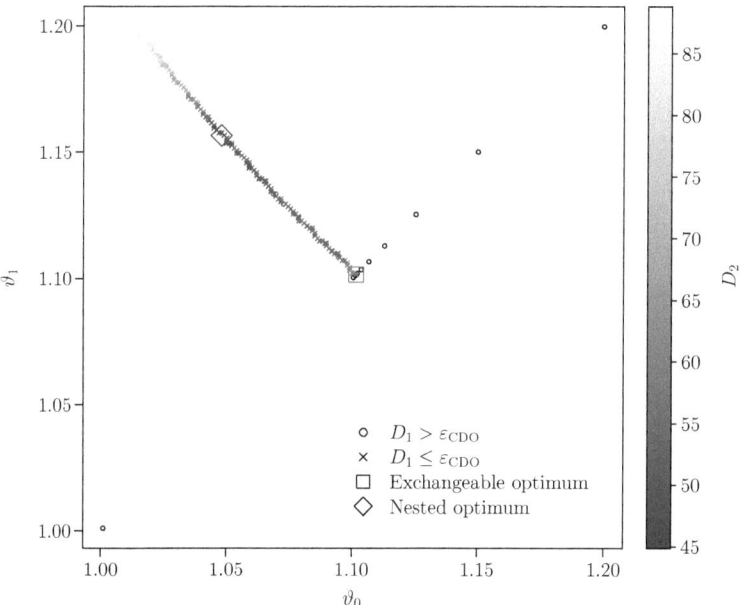

Figure 5.4 Progression of the CDO calibration algorithm for the outer power Clayton copula (2007-06-12, $T = 10$). The crosses and circles indicate whether the upfront criterion $D_1 \leq \varepsilon_{\text{CDO}}$ is met or not, respectively. The error D_2, illustrated with different shades of gray, is optimized over all marked points in the $\vartheta_0 \vartheta_1$-plane satisfying $D_1 \leq \varepsilon_{\text{CDO}}$. The square and the diamond indicate the minimizing arguments for the Archimedean and nested Archimedean copula, respectively.

5.5 Conclusion

We introduced the class of nested Archimedean copulas to the copula approach of Li (2000) and Schönbucher and Schubert (2001) for the modeling of dependent defaults. This class of copulas induces a hierarchical structure on the obligors in the considered credit portfolio, which, depending on the classification criterion, allows for different economic interpretations. To demonstrate the advantage of using nested Archimedean copulas over their exchangeable counterparts, we calibrated the model to CDO tranche spreads of the Markit iTraxx Europe indices. The hierarchical structure for this calibration was defined by the original iTraxx industry sector segmentation. Our analysis includes several Archimedean families, some of which were newly applied within this framework, and therefore indicates which copulas might be preferable for modeling CDOs. As a benchmark for the conducted calibration, we included the Gaussian copula to represent the current standard market model. The calibration results show considerably reduced pricing errors by using nested Archimedean copulas, even if we restrict our model to identical sector parameters. Moreover, our results indicate that copulas which are able to capture upper tail dependence provide the most accurate fits, e.g., Gumbel's, Joe's and the outer power Clayton family performed the best. Technically, such a calibration requires fast simulation techniques and an optimizer which exploits the specific structure of the problem. The former were presented in Chapter 4, the latter was introduced in this chapter. Further results addressed confidence intervals for CDO tranche spreads and the implied term structure of default correlations of the model. Additionally, we showed that our generalization allows for larger default correlations among firms in the same sector compared to firms in different sectors.

A Supplementary material

In the following three sections we recall the notion of conditional distribution functions, briefly present Marshall-Olkin copulas, and give an introduction to Laplace and Laplace-Stieltjes transforms.

A.1 Conditional distribution functions

Conditional distribution functions play a role in the conditional distribution method for generating random variates from a given copula. In this section, we briefly recall the notions of conditional expectations, probabilities, and distribution functions.

Definition A.1.1

If $(\Omega, \mathcal{F}, \mathbb{P})$ is a probability space, $X : \Omega \to \mathbb{R}$ is $\mathcal{F}$-measurable with $\mathbb{E}|X| < \infty$, and $\mathcal{G} \subseteq \mathcal{F}$ is a σ-algebra, then any random variable Y, such that

(1) Y is $\mathcal{G}$-measurable,

(2) $\mathbb{E}|Y| < \infty$,

(3) $\int_G Y \, d\mathbb{P} = \int_G X \, d\mathbb{P}$ for all $G \in \mathcal{G}$,

is called a *conditional expectation of X given $\mathcal{G}$*, denoted by $\mathbb{E}[X|\mathcal{G}]$. Further, for $A \in \mathcal{F}$, $\mathbb{P}(A|\mathcal{G}) := \mathbb{E}[\mathbb{1}_A|\mathcal{G}]$ is called a *conditional probability of A given $\mathcal{G}$*.

For a proof that such a Y exists, is indeed integrable, and almost surely unique, therefore also referred to as a *version*, see, e.g., Durrett (2004, pp. 217). If $\mathcal{G}$ is a σ-algebra generated by at most countably many sets which build a partition on Ω and each of which has positive measure, then elementary conditional expectations and probabilities arise as special cases.

If $(S, \mathcal{S})$ is a measurable space and $X : \Omega \to S$ is $(\mathcal{F}, \mathcal{S})$-measurable, then $\mathbb{P}(X \in \cdot|\mathcal{G})$ is not necessarily a probability measure on $(S, \mathcal{S})$ for a.e. $\omega \in \Omega$, see Doob (1990, p. 624) for a counterexample. This gives rise to the following definition.

A Supplementary material

Definition A.1.2

Let $(\Omega, \mathcal{F}, \mathbb{P})$ be a probability space, $X : \Omega \to S$ $(\mathcal{F}, \mathcal{S})$-measurable, and $\mathcal{G} \subseteq \mathcal{F}$ a σ-algebra. A mapping $\mathbb{P}^* : \mathcal{S} \times \Omega \to [0,1]$, such that

(1) for all $A \in \mathcal{S}$, $\mathbb{P}^*(A, \cdot)$ is a version of $\mathbb{P}(X \in A | \mathcal{G})$;

(2) for a.e. $\omega \in \Omega$, $\mathbb{P}^*(\cdot, \omega)$ is a probability measure on $(S, \mathcal{S})$,

is called a *regular conditional distribution*.

If $(S, \mathcal{S})$ is a standard Borel space, e.g., a Polish space, see Klenke (2008, p. 185), then a regular version of $\mathbb{P}(X \in \cdot | \mathcal{G})$ exists and in the sequel we assume $\mathbb{P}(X \in \cdot | \mathcal{G})$ to be such a version.

One often considers the case that $\mathcal{G}$ is the σ-algebra generated by some $(\mathcal{F}, \mathcal{S}')$-measurable random vector $\boldsymbol{Y} : \Omega \to S'$, where $(S', \mathcal{S}')$ is another measurable space. In this case define $\mathbb{E}[X|\boldsymbol{Y}] := \mathbb{E}[X|\sigma(\boldsymbol{Y})]$, and analogously for $\mathbb{P}(A|\boldsymbol{Y})$. By the Factorization Lemma, see Bauer (1992, pp. 71), $\mathbb{E}[X|\boldsymbol{Y}] = g(\boldsymbol{Y})$ for an $\mathcal{S}'$-measurable $g : S' \to \mathbb{R}$. Thus, define the *conditional expectation of X given $\boldsymbol{Y} = \boldsymbol{y}$* by

$$\mathbb{E}[X|\boldsymbol{Y} = \boldsymbol{y}] := g(\boldsymbol{y}), \ \boldsymbol{y} \in S'.$$

Similarly for the *conditional probability of A given $\boldsymbol{Y} = \boldsymbol{y}$*, denoted by $\mathbb{P}[A|\boldsymbol{Y} = \boldsymbol{y}]$.

For two random vectors $\boldsymbol{X} : \Omega \to \mathbb{R}^n$, $\boldsymbol{Y} : \Omega \to \mathbb{R}^m$, one can therefore define the *conditional distribution function* $F_{\boldsymbol{X}|\boldsymbol{Y}}(\boldsymbol{x}|\boldsymbol{y}) := \mathbb{P}(\boldsymbol{X} \leq \boldsymbol{x} | \boldsymbol{Y} = \boldsymbol{y})$. This implies that $F_{\boldsymbol{X}|\boldsymbol{Y}}(\boldsymbol{x}|\boldsymbol{y}) = \mathbb{E}[\mathbb{1}_{\{\boldsymbol{X} \leq \boldsymbol{x}\}}|\boldsymbol{Y} = \boldsymbol{y}]$. It follows that $\int_{(-\infty, \boldsymbol{y}]} F_{\boldsymbol{X}|\boldsymbol{Y}}(\boldsymbol{x}|\boldsymbol{y}')\,dF_{\boldsymbol{Y}}(\boldsymbol{y}') = \int_{(-\infty, \boldsymbol{y}]} \mathbb{E}[\mathbb{1}_{\{\boldsymbol{X} \leq \boldsymbol{x}\}}|\boldsymbol{Y} = \boldsymbol{y}']\,dF_{\boldsymbol{Y}}(\boldsymbol{y}') = \mathbb{E}[\mathbb{1}_{\{\boldsymbol{Y} \leq \boldsymbol{y}\}}\mathbb{E}[\mathbb{1}_{\{\boldsymbol{X} \leq \boldsymbol{x}\}}|\boldsymbol{Y}]] = \mathbb{E}[\mathbb{E}[\mathbb{1}_{\{\boldsymbol{X} \leq \boldsymbol{x}, \boldsymbol{Y} \leq \boldsymbol{y}\}}|\boldsymbol{Y}]]$. By the tower property for conditional expectations, see Williams (1991, p. 88),

$$F_{\boldsymbol{X}, \boldsymbol{Y}}(\boldsymbol{x}, \boldsymbol{y}) := \mathbb{P}(\boldsymbol{X} \leq \boldsymbol{x}, \boldsymbol{Y} \leq \boldsymbol{y}) = \int_{(-\infty, \boldsymbol{y}]} F_{\boldsymbol{X}|\boldsymbol{Y}}(\boldsymbol{x}|\boldsymbol{y}')\,dF_{\boldsymbol{Y}}(\boldsymbol{y}'), \ \boldsymbol{x} \in \mathbb{R}^n, \ \boldsymbol{y} \in \mathbb{R}^m. \quad (A.1)$$

A.2 Marshall-Olkin copulas

Marshall and Olkin (1967) presented a multivariate exponential distribution. The class of Marshall-Olkin copulas which corresponds to this distribution via Sklar's Theorem is best interpreted in two dimensions. Consider a system of two components which are affected by two fatal shocks hitting the components individually and one shock hitting both components simultaneously. Assuming that the shocks follow three independent Poisson processes, the times of occurrence of the three shocks are given by independent exponential random variables

A.2 Marshall-Olkin copulas

$Z_1 \sim \text{Exp}(\lambda_1)$, $Z_2 \sim \text{Exp}(\lambda_2)$, and $Z_{12} \sim \text{Exp}(\lambda_{12})$, respectively. Therefore, the lifetime of component j of the system is given by $X_j := \min\{Z_j, Z_{12}\}$, $j \in \{1,2\}$. By introducing the parameters $\alpha_j := \lambda_{12}/(\lambda_j + \lambda_{12}) \in (0,1)$, the survival copula $C_{\alpha_1\alpha_2}$ corresponding to the vector of lifetimes $(X_1, X_2)^T$ is the bivariate *Marshall-Olkin copula*, given by

$$C_{\alpha_1\alpha_2}(u_1, u_2) := \min\{u_1^{1-\alpha_1} u_2, u_1 u_2^{1-\alpha_2}\}.$$

The range of possible parameters can be extended to the case where $\alpha_j \in [0,1]$, $j \in \{1,2\}$, with $C_{0\alpha_2} = C_{\alpha_1 0} = \Pi$ and $C_{11} = M$. The case $\alpha_1 = \alpha_2$ is referred to as bivariate *Cuadras-Augé copula*, see Cuadras and Augé (1981).

A random vector $(U_1, U_2)^T \sim C_{\alpha_1\alpha_2}$ allows for the stochastic representation

$$U_1 = \bar{F}_1(X_1) = \exp(-(\lambda_1 + \lambda_{12})\min\{Z_1, Z_{12}\}),$$
$$U_2 = \bar{F}_2(X_2) = \exp(-(\lambda_2 + \lambda_{12})\min\{Z_2, Z_{12}\}), \tag{A.2}$$

which can be applied for sampling a bivariate Marshall-Olkin copula.

Marshall-Olkin copulas have both an absolutely continuous and a singular component. The computations are based on the fact that

$$\mathrm{D}_{21} C_{\alpha_1\alpha_2}(u_1, u_2) = \begin{cases} (1-\alpha_1) u_1^{-\alpha_1}, & u_1^{\alpha_1} < u_2^{\alpha_2}, \\ (1-\alpha_2) u_2^{-\alpha_2}, & u_1^{\alpha_1} > u_2^{\alpha_2} \end{cases} \tag{A.3}$$

and carried out, e.g., in Nelsen (2007, p. 54). The mass on the singular component equals $\mathbb{P}(U_1^{\alpha_1} = U_2^{\alpha_2}) = \alpha_1\alpha_2/(\alpha_1 + \alpha_2 - \alpha_1\alpha_2)$.

For bivariate Marshall-Olkin copulas, the measures of concordance Spearman's rho and Kendall's tau can be computed via Theorem 1.7.8 and are given by

$$\rho_{S, C_{\alpha_1\alpha_2}} = \frac{3\alpha_1\alpha_2}{2(\alpha_1+\alpha_2) - \alpha_1\alpha_2} \quad \text{and} \quad \tau_{C_{\alpha_1\alpha_2}} = \frac{\alpha_1\alpha_2}{\alpha_1 + \alpha_2 - \alpha_1\alpha_2},$$

where the latter formula is obtained by an application of Identity (1.8). The lower and upper tail-dependence coefficient can be found via Theorem 1.7.11 (1), (5) and are given by

$$\lambda_l = 0 \text{ and } \lambda_u = \min\{\alpha_1, \alpha_2\}.$$

Bivariate Marshall-Olkin copulas extend to more than two dimensions, leading to a multivariate copula with $\sum_{j=2}^{d} \binom{d}{j} j$ parameters in d dimensions. Although the construction principle via lifetimes of components in a system affected by shocks extends to the multivariate case, Representation (A.2) gets notationally complicated. For more details about the d-dimensional case, see Li (2008). As the bivariate margins are again Marshall-Olkin copulas, the formulas for Spearman's rho, Kendall's tau, and the tail-dependence coefficients carry over to the general case.

A Supplementary material

A.3 Laplace-Stieltjes transforms

In this section, we give an introduction to Laplace-Stieltjes transforms. To understand this concept, we first give a brief overview of Lebesgue-Stieltjes and Riemann-Stieltjes integrals. For more details, we refer the reader to Apostol (1974), Widder (1946), Arendt et al. (2002), and Folland (1999). In the sequel, $\mathbb{K}$ stands for either $\mathbb{R}$ or $\mathbb{C}$.

A function $f : [a,b] \to \mathbb{K}$, $a,b \in \mathbb{R} : a < b$, is *of bounded variation on* $[a,b]$, if $V_{[a,b]}(f) < \infty$, where

$$V_{[a,b]}(f) := \sup \left\{ \sum_{k=1}^{n} |f(x_k) - f(x_{k-1})| \, : \, n \in \mathbb{N}, a =: x_0 < \cdots < x_n := b \right\},$$

denoted by $f \in BV([a,b]) := \{f : [a,b] \to \mathbb{K} \, : \, V_{[a,b]}(f) < \infty\}$. Let $a, b \in \mathbb{R} : a < b$ and $g : [a,b] \to \mathbb{K}$, and consider an integral of the form

$$\int_{[a,b]} g(x) \, dF(x). \tag{A.4}$$

First consider $\mathbb{K} = \mathbb{R}$ and let $F : [a,b] \to \mathbb{R}$ be bounded and increasing. Now extend F to $F : \mathbb{R} \to \mathbb{R}$ with $F(x) := a$ for $x < a$ and $F(x) := b$ for $x > b$ and define $F^* : \mathbb{R} \to \mathbb{R}$ by $F^*(x) := F(x+) - F(a)$. Then F^* is bounded, increasing, and right-continuous. Thus, Folland (1999, p. 35) implies that there is a unique finite and complete Borel measure μ_{F^*} on $\mathbb{R}$, such that $\mu_{F^*}((c,d]) = F^*(d) - F^*(c)$ for all $c, d \in \mathbb{R} : c < d$, called the *Lebesgue-Stieltjes measure* associated to F^*. Thus, for bounded and increasing functions $F : [a,b] \to \mathbb{R}$, (A.4) is defined by the Lebesgue integral with respect to μ_{F^*}, i.e.,

$$\int_{[a,b]} g(x) \, dF(x) := \int_{[a,b]} g(x) \, d\mu_{F^*}(x).$$

Now if $\mathbb{K} = \mathbb{R}$ and $F \in BV([a,b])$, F can be decomposed into $F(x) = \frac{1}{2}(V_{[a,x]}(f) + F(x)) - \frac{1}{2}(V_{[a,x]}(f) - F(x)) =: F_1(x) - F_2(x)$, where F_1, F_2 are two non-negative, bounded, and increasing functions. Integral (A.4) with integrator F is then defined as the difference of the corresponding Lebesgue-Stieltjes integrals with integrators F_1 and F_2. Finally, if $\mathbb{K} = \mathbb{C}$ and $F \in BV([a,b])$, Integral (A.4) is defined via the corresponding real and imaginary part. If Integral (A.4) defined this way exists, it is called the *Lebesgue-Stieltjes integral* and g is said to be *Lebesgue-Stieltjes integrable with respect to* F. This notion generalizes to improper integrals, defined as usual via the corresponding limit, if it exists. If so, this definition generalizes the Lebesgue-Stieltjes integral viewed as a Lebesgue integral on an unbounded interval, for, if the latter exists, it must coincide with the improper Lebesgue-Stieltjes integral by the Dominated Convergence Theorem, see Folland (1999, p. 54).

A.3 Laplace-Stieltjes transforms

Now let us again consider $\mathbb{K} = \mathbb{R}$, $F : [a,b] \to \mathbb{R}$ to be bounded and increasing, and F^* to be its standardized, bounded, increasing, and right-continuous analog as before. By an extension of Lebesgue's Integrability Criterion to Stieltjes integrals, see Elstrodt (2005, p. 153), the existence of Integral (A.4) as a Riemann-Stieltjes integral, see Apostol (1974) or Arendt et al. (2002) for more details on such integrals, implies the existence of Integral (A.4) as a Lebesgue-Stieltjes integral and both are equal. By the same argument as before, this generalizes to the case where $F \in \mathrm{BV}([a,b])$. Thus, Lebesgue-Stieltjes integrals are generalizations of Riemann-Stieltjes integrals. If Integral (A.4) exists in the Riemann-Stieltjes sense, g is said to be *Riemann-Stieltjes integrable with respect to F*.

Prominent sufficient conditions for the existence of Riemann-Stieltjes integrals of Type (A.4) are if g is continuous on $[a,b]$ and $F \in \mathrm{BV}([a,b])$, or if F is continuous on $[a,b]$ and $g \in \mathrm{BV}([a,b])$, see Apostol (1974, p. 159), or if $g \in \mathrm{BV}([a,b])$, $F \in \mathrm{BV}([a,b])$, and g and F have no common points of discontinuities, see Widder (1946, p. 25). If g is Riemann-Stieltjes integrable with respect to F, the second of these conditions follows from the first by interchanging the role of g and F via the *Integration by Parts Formula*

$$\int_a^b g(x)\, dF(x) = g(b)F(b) - g(a)F(a) - \int_a^b F(x)\, dg(x),$$

see Apostol (1974, p. 144) or Arendt et al. (2002, p. 52). Besides the Integration by Parts Formula, important properties of Riemann-Stieltjes integrals include linearity and additivity, see Apostol (1974, pp. 142), and the change of variable formulas, see Apostol (1974, pp. 144). For Riemann-Stieltjes integrals depending on a parameter, differentiation under the integral sign, or interchanging the order of integration are important properties, see Apostol (1974, pp. 166).

There are two obvious special cases if one considers Riemann-Stieltjes integrals of Type (A.4). The first one is obtained if F is a *step function*, i.e., a function which has at most finitely many jumps, assumed to be at $x_1, \ldots, x_n$ with heights $y_1, \ldots, y_n$, respectively, and is constant in between. If $g : [a,b] \to \mathbb{K}$ is a Riemann integrable function such that g and F are not both discontinuous from the left or from the right at x_k for each $k \in \{1, \ldots, n\}$, then Integral (A.4) exists and

$$\int_a^b g(x)\, dF(x) = \sum_{k=1}^n g(x_k) y_k,$$

see Apostol (1974, pp. 148). The second special case is obtained from Lebesgue-Stieltjes integrals if $g : [a,b] \to \mathbb{K}$ is continuous and $F : [a,b] \to \mathbb{K}$ is *absolutely continuous* on $[a,b]$, i.e., for all $\varepsilon > 0$ there exists a $\delta > 0$ such that for any finite set of pairwise disjoint intervals (x_k, y_k), $k \in \{1, \ldots, n\}$, $n \in \mathbb{N}$, with $\sum_{k=1}^n |y_k - x_k| < \delta$ it holds that $\sum_{k=1}^n |F(y_k) - F(x_k)| < \varepsilon$. By the

A Supplementary material

Fundamental Theorem of Calculus, see Folland (1999, p. 106), $F(x) = F(a) + \int_a^x f(t)\,dt$, where f denotes the almost everywhere existing and Lebesgue integrable derivative F' of F. This implies $F \in \mathrm{BV}([a,b])$, see Folland (1999, p. 106). Thus, Integral (A.4) exists and

$$\int_a^b g(x)\,dF(x) = \int_a^b g(x)f(x)\,dx, \tag{A.5}$$

see Arendt et al. (2002, p. 55).

With these concepts at hand, let us now turn to Laplace-Stieltjes and Laplace transforms. References include the textbooks of Widder (1946), Doetsch (1970), and Arendt et al. (2002), the latter even dealing with a more general setup.

Definition A.3.1
For $F \in \mathrm{BV}_{\mathrm{loc}}([0,\infty)) := \{f : [0,\infty) \to \mathbb{K} : \mathrm{V}_{[0,r]}(f) < \infty \text{ for all } r > 0\}$, the *Laplace-Stieltjes transform of F* is defined by the improper Lebesgue-Stieltjes integral

$$\mathcal{LS}[F](s) := \int_0^\infty \exp(-sx)\,dF(x) := \lim_{r \to \infty} \int_{[0,r]} \exp(-sx)\,dF(x), \tag{A.6}$$

for all $s \in \mathbb{C}$ where the limit exists. The extended real number $\mathrm{abs}(\mathcal{LS}[F]) := \inf\{\mathrm{Re}\,s : \mathcal{LS}[F](s) \text{ exists}\}$ is called the *abscissa of convergence of $\mathcal{LS}[F]$*. If $\mathrm{abs}(\mathcal{LS}[F]) < \infty$, then F is called *Laplace-Stieltjes transformable*.

Continuity of $x \mapsto \exp(-sx)$ for every $s \in \mathbb{C}$ readily implies the existence of the integrals $\int_{[0,r]} \exp(-sx)\,dF(x)$, $r > 0$, on the right-hand side of Definition (A.6), in both the Riemann-Stieltjes and the Lebesgue-Stieltjes sense. Further, Laplace-Stieltjes transforms of monotone and bounded functions F on $[0,\infty)$ exist for all $s \in \mathbb{C} : \mathrm{Re}\,s > 0$. For more general results about the existence of Laplace-Stieltjes transforms, see, e.g., Widder (1946, pp. 38), who also presents formulas for the abscissa of convergence.

The following result establishes the differentiability of Laplace-Stieltjes transforms. It may be found in Arendt et al. (2002, pp. 58).

Theorem A.3.2 (Frequency Differentiation Theorem)
Let $F \in \mathrm{BV}_{\mathrm{loc}}([0,\infty))$. Then $\mathcal{LS}[F](s)$ converges for all $s \in \mathbb{C} : \mathrm{Re}\,s > \mathrm{abs}(\mathcal{LS}[F])$ and diverges for all $s \in \mathbb{C} : \mathrm{Re}\,s < \mathrm{abs}(\mathcal{LS}[F])$. Further, if F is Laplace-Stieltjes transformable, then $\mathcal{LS}[F](s)$ is holomorphic for all $s \in \mathbb{C} : \mathrm{Re}\,s > \mathrm{abs}(\mathcal{LS}[F])$ and the k-th derivative of $\mathcal{LS}[F]$, $k \in \mathbb{N}_0$, is given by

$$\mathcal{LS}[F]^{(k)}(s) = \int_0^\infty (-x)^k \exp(-sx)\,dF(x), \ s \in \mathbb{C} : \mathrm{Re}\,s > \mathrm{abs}(\mathcal{LS}[F]).$$

A.3 Laplace-Stieltjes transforms

If $F(x) = \int_0^x f(t)\,dt$ for $f \in L^1_{\text{loc}}([0,\infty)) := \{f : [0,\infty) \to \mathbb{K} : f \text{ is Lebesgue integrable on } [0,r]$ for all $r > 0\}$, applying Formula (A.5) leads to

$$\int_0^r \exp(-sx)\,dF(x) = \int_0^r \exp(-sx) f(x)\,dx, \qquad (A.7)$$

so that $\mathcal{LS}[F](s)$ exists if and only if the limit on the right-hand side of (A.7) exists, in which case both are equal, see Arendt et al. (2002, p. 56). This gives rise to the following definition.

Definition A.3.3

For $f \in L^1_{\text{loc}}([0,\infty))$, the *Laplace transform of f* is defined by the improper Lebesgue integral

$$\mathcal{L}[f](s) := \int_0^\infty \exp(-sx) f(x)\,dx := \lim_{r \to \infty} \int_{[0,r]} \exp(-sx) f(x)\,dx,$$

for all $s \in \mathbb{C}$ where the limit exists. The extended real number $\text{abs}(\mathcal{L}[f]) := \inf\{\text{Re}\,s : \mathcal{L}[f](s) \text{ exists}\}$ is called the *abscissa of convergence of $\mathcal{L}[f]$*. If $\text{abs}(\mathcal{L}[f]) < \infty$, then f is said to be *Laplace transformable*.

The Laplace transform is a special case of the Laplace-Stieltjes transform. Similar to the latter, $\mathcal{L}[f](s)$ converges for all $\text{Re}\,s > \text{abs}(\mathcal{L}[f])$ and diverges for all $s \in \mathbb{C} : \text{Re}\,s < \text{abs}(\mathcal{L}[f])$, see Arendt et al. (2002, p. 28). If $\text{abs}(\mathcal{L}[f]) < \infty$, $\mathcal{L}[f](s)$ is holomorphic for all $s \in \mathbb{C} : \text{Re}\,s > \text{abs}(\mathcal{L}[f])$ and the k-th derivative of $\mathcal{L}[f]$, $k \in \mathbb{N}_0$, is given by

$$\mathcal{L}[f]^{(k)}(s) = \int_0^\infty (-x)^k \exp(-sx) f(x)\,dx, \ s \in \mathbb{C} : \text{Re}\,s > \text{abs}(\mathcal{L}[f]),$$

see Arendt et al. (2002, p. 33). The Integration by Parts Formula can be used to establish the general connection

$$\int_0^r \exp(-sx)\,dF(x) = \exp(-sr) F(r) - F(0) + s \int_0^r \exp(-sx) F(x)\,dx, \qquad (A.8)$$

see Arendt et al. (2002, p. 56). Equation (A.8) is important in that it allows one to transfer results from Laplace to Laplace-Stieltjes transforms and vice versa. One such result is the following, see Widder (1946, p. 41) and Arendt et al. (2002, p. 39).

Theorem A.3.4 (Integration and Differentiation Theorem)

If $\mathcal{LS}[F](s)$ exists for some $s \in \mathbb{C} : \text{Re}\,s > 0$, then $\mathcal{L}[F](s)$ exists and

$$\mathcal{LS}[F](s) = s\mathcal{L}[F](s) - F(0).$$

If F is absolutely continuous and $\mathcal{L}[F'](s)$ exists for some $s \in \mathbb{C} : \text{Re}\,s > 0$, then $\mathcal{LS}[F](s)$ exists and equals $\mathcal{L}[F'](s)$ so that $\mathcal{L}[F'](s) = s\mathcal{L}[F](s) - F(0)$.

A Supplementary material

Note that the second statement of the Integration and Differentiation Theorem follows from the first by applying the Fundamental Theorem of Calculus, see Folland (1999, p. 106), and using Identity (A.7).

Concerning the uniqueness of Laplace-Stieltjes and Laplace transforms, see the following result. A proof for the first statement may be found in Doetsch (1970, p. 35), the second statement follows from the Integration and Differentiation Theorem.

Theorem A.3.5 (Uniqueness Theorem)
If $\mathrm{abs}(\mathcal{L}[f]) < \infty$ and $\mathcal{L}[f](s + nh) = 0$ for some $s \in \mathbb{C} : \mathrm{Re}\, s > \mathrm{abs}(\mathcal{L}[f])$, some $h > 0$, and all $n \in \mathbb{N}_0$, then $f = 0$ almost everywhere. If $\mathrm{abs}(\mathcal{LS}[F]) < \infty$, $F(0) = 0$, and $\mathcal{L}[F](s + nh) = 0$ for some $s \in \mathbb{C} : \mathrm{Re}\, s > \mathrm{abs}(\mathcal{LS}[F])$, some $h > 0$, and all $n \in \mathbb{N}_0$, then $F = 0$ almost everywhere.

By taking $f := f_1 - f_2$, respectively $F := F_1 - F_2$ with $F_1(0) = F_2(0)$, in the Uniqueness Theorem, and by using linearity of $\mathcal{L}[f]$ and $\mathcal{LS}[F]$, one readily obtains that $f_1 = f_2$ and $F_1 = F_2$ almost everywhere. If one additionally has, e.g., right-continuity, then the functions are equal everywhere. We frequently apply the Uniqueness Theorem to the setup where f and F are right-continuous and $F(0) = 0$. The corresponding inverse transform of the Laplace transform is denoted by $\mathcal{L}^{-1}$, the one for the Laplace-Stieltjes transform by $\mathcal{LS}^{-1}$.

There are several formulas available for constructing f from its Laplace transform $\mathcal{L}[f]$, two of which are briefly discussed in the sequel. For the following result, see, e.g., Widder (1946, p. 66). As usual, $\mathcal{L}[f](s)$ is called *absolutely convergent* if $\lim_{r \to \infty} \int_{[0,r]} |\exp(-sx) f(x)|\, dx$ exists.

Theorem A.3.6 (Bromwich integral inversion)
Let $f \in \mathrm{L}^1_{\mathrm{loc}}((0, \infty))$ and $c \in \mathbb{R}$ such that $\mathcal{L}[f](s)$ converges absolutely for all $s \in \mathbb{C} : \mathrm{Re}\, s = c$. Further, let $x \in \mathbb{R}$ and assume for $x \in [0, \infty)$ that f is of bounded variation in a neighborhood of x, being a right-hand neighborhood if $x = 0$. Then, the *Bromwich integral* is given by

$$\lim_{r \to \infty} \frac{1}{2\pi i} \int_{c-ir}^{c+ir} \exp(sx) \mathcal{L}[f](s)\, ds = \begin{cases} 0, & x < 0, \\ f(x+)/2, & x = 0, \\ (f(x-) + f(x+))/2, & x > 0. \end{cases}$$

Theorem A.3.6 is often applied in the setup as given in the following corollary. Its proof is straightforward by applying the second statement of the Integration and Differentiation Theorem.

Corollary A.3.7
If F is an absolutely continuous distribution function on $[0, \infty)$ with density f, $F(0) = 0$, and

$c \in (0, \infty)$, then
$$\lim_{r \to \infty} \frac{1}{2\pi i} \int_{c-ir}^{c+ir} \exp(sx) \frac{\mathcal{LS}[F](s)}{s} ds = F(x), \quad x \in [0, \infty).$$

We mainly consider Laplace and Laplace-Stieltjes transforms as functions on the non-negative real line and write t instead of s in this case. Further, $t = \infty$ denotes the corresponding limit as usual.

An inversion formula for Laplace transform using only real values of the transform was found by Post (1930), see Widder (1946, p. 289). It is therefore known as *Post's Formula*.

Theorem A.3.8 (Post (1930))

Let $f \in \mathrm{BV}_{\mathrm{loc}}([0, \infty))$ such that $\mathcal{L}[f](t)$ converges absolutely for some $t \in \mathbb{R}$. Then
$$\lim_{n \to \infty} \frac{(-1)^n}{n!} \left(\frac{n}{x}\right)^{n+1} \mathcal{L}[f]^{(n)}\left(\frac{n}{x}\right) = \frac{f(x-) + f(x+)}{2}, \quad x \in (0, \infty).$$

Again applying the second statement of the Integration and Differentiation Theorem leads to the following result.

Corollary A.3.9

If F is an absolutely continuous distribution function on $[0, \infty)$ with density f and $F(0) = 0$, then
$$\lim_{n \to \infty} \frac{(-1)^n}{n!} \left(\frac{n}{x}\right)^{n+1} \left(\frac{\mathcal{LS}[F](s)}{s}\right)^{(n)}\left(\frac{n}{x}\right) = F(x), \quad x \in (0, \infty).$$

In Subsection 2.1.2 we need the following continuity theorem, see Feller (1971, p. 431) and Feller (1971, p. 433) for a proof.

Theorem A.3.10 (Continuity)

For $n \in \mathbb{N}$, let F_n be the distribution function of the finite measure μ_n on $[0, \infty)$, i.e., $F_n(x) := \mu_n([0, x])$, $x \in [0, \infty)$, and let $\psi_n := \mathcal{LS}[F_n]$.

(1) Let F be a distribution function of a finite measure μ on $[0, \infty)$ with $\psi := \mathcal{LS}[F]$. If μ_n converges weakly to μ for $n \to \infty$, equivalently $\lim_{n \to \infty} F_n(x) = F(x)$ at all continuity points x of F, and if $(\psi_n(0))_{n \in \mathbb{N}}$ is bounded, then $\lim_{n \to \infty} \psi_n(t) = \psi(t)$ for all $t \in (0, \infty)$.

(2) If $\lim_{n \to \infty} \psi_n(t) = \psi(t)$ for all $t \in (0, \infty)$, then $\psi = \mathcal{LS}[F]$, where F is the distribution function of a finite measure μ on $[0, \infty)$, and μ_n converges weakly to μ. In the case where μ_n is a probability measure for all $n \in \mathbb{N}$, μ is possibly a *defective* probability measure, i.e., $\mu([0, \infty)) < 1$, and $\mu([0, \infty)) = 1$ if and only if $\psi(0) = 1$.

A Supplementary material

We also frequently use the following result, which is straightforward to verify.

Proposition A.3.11 (Convolutions)
Let $X_i \sim F_i$, $i \in \{1, \ldots, n\}$, be independent random variables on $[0, \infty)$ and let G be the distribution function of $\sum_{i=1}^{n} X_i$. Then,

$$\mathcal{LS}[G](s) = \prod_{i=1}^{n} \mathcal{LS}[F_i](s), \ s \in \mathbb{C} : \operatorname{Re} s > 0.$$

If, additionally, X_i, $i \in \{1, \ldots, n\}$, are identically distributed according to the distribution function F, then $\mathcal{LS}[G](s) = \mathcal{LS}[F]^n(s)$ for all $s \in \mathbb{C} : \operatorname{Re} s > 0$.

B Supplementary CDO pricing results

B Supplementary CDO pricing results

2007-06-14	Dependence		CDO upfront and spreads $\hat{s}_{5,j}^{\mathrm{CDO},f}$					Error
Copula	$\hat{\vartheta}$	ρ in %	$j=1$	$j=2$	$j=3$	$j=4$	$j=5$	D_2
A	– – –	– – –	– –	– –	– –	– –	– –	– –
$\hat{A}$	0.68 0.18 0.93	3.42 0.37 16.09	7.33 7.38	111.15 97.68	1.63 8.89	0.00 0.54	0.00 0.01	83.07 61.78
C	1.84 1.71 1.97	3.11 2.90 3.33	7.33 7.40	92.13 88.64	16.68 15.13	2.92 2.66	0.17 0.14	55.27 50.52
$\hat{C}$	0.08 0.08 0.08	3.01 2.90 3.13	7.35 7.40	84.72 81.94	18.09 17.07	4.50 4.33	0.43 0.46	47.43 43.76
opC	1.06 1.03 1.09	7.41 4.43 10.98	7.36 7.38	34.38 45.81	16.31 19.42	10.88 10.29	6.31 4.12	25.58 14.60
$\hat{\mathrm{opC}}$	– – –	– – –	– –	– –	– –	– –	– –	– –
F	2.64 2.53 2.74	3.05 2.90 3.18	7.34 7.36	86.59 82.77	19.47 17.58	4.71 4.00	0.49 0.40	50.41 45.50
G	1.06 1.03 1.09	7.46 4.13 11.33	7.36 7.40	32.68 44.76	15.82 19.56	10.48 10.17	6.07 3.89	26.14 15.32
$\hat{G}$	1.23 1.23 1.24	3.01 2.95 3.14	7.34 7.37	99.11 97.29	9.97 9.58	0.52 0.51	0.00 0.00	62.16 60.74
J	1.06 1.03 1.10	7.64 3.90 12.02	7.36 7.36	29.57 45.47	15.63 20.76	10.54 10.66	6.19 3.87	29.23 16.27
$\hat{J}$	– – –	– – –	– –	– –	– –	– –	– –	– –
Ga	0.19 0.18 0.20	3.05 2.89 3.38	7.35 7.40	83.83 81.26	18.67 17.67	5.03 4.75	0.60 0.61	46.42 43.11
Market			7.36	45.75	12.00	5.11	2.19	

Table B.1 Results as at 2007-06-14 based on 500 000 runs and $T = 5$.

2007-06-14	Dependence		CDO upfront and spreads $\hat{s}_{10,j}^{\text{CDO},f}$					Error
Copula	$\hat{\vartheta}$	ρ in %	$j=1$	$j=2$	$j=3$	$j=4$	$j=5$	D_2
A	0.73	4.88	35.78	361.29	148.29	61.75	11.43	135.42
	0.68 0.92	4.54 6.12	35.80	359.26	146.44	60.22	10.99	130.46
$\hat{\text{A}}$	0.29	2.51	35.73	525.84	105.94	3.98	0.01	285.27
	0.01 0.84	0.04 20.48	35.75	417.88	147.00	41.44	3.13	178.71
C	0.66	4.48	35.73	379.31	149.00	56.51	8.63	151.71
	0.63 0.72	4.28 4.85	35.77	378.51	147.46	55.00	8.05	148.44
$\hat{\text{C}}$	0.07	4.17	35.73	379.94	143.58	53.54	8.66	143.93
	0.07 0.08	4.13 4.25	35.78	379.35	142.07	52.70	8.60	141.05
opC	1.10	12.86	35.79	269.11	87.50	47.31	24.73	57.96
	1.05 1.16	6.45 18.65	35.78	306.55	114.26	57.17	18.79	47.22
o$\hat{\text{p}}$C	–	–	–	–	–	–	–	–
	– –	– –	–	–	–	–	–	–
F	1.60	6.16	35.79	320.40	140.91	69.88	19.95	101.59
	1.58 1.61	6.09 6.23	35.74	318.45	139.24	68.49	19.15	95.79
G	1.13	14.55	35.73	239.22	84.81	50.83	27.88	97.20
	1.06 1.20	7.08 21.10	35.80	285.84	112.16	60.24	21.19	70.98
$\hat{\text{G}}$	1.11	3.08	35.73	455.02	144.26	22.87	0.45	233.43
	1.10 1.16	2.86 4.64	35.75	445.22	144.11	26.71	0.74	219.35
J	1.17	16.81	35.73	204.43	75.27	50.33	30.29	143.44
	1.03 1.29	3.21 25.71	35.79	301.25	129.48	68.99	17.32	77.77
$\hat{\text{J}}$	–	–	–	–	–	–	–	–
	– –	– –	–	–	–	–	–	–
Ga	0.14	4.50	35.75	371.55	143.90	56.63	10.23	137.37
	0.13 0.17	4.29 5.51	35.73	371.40	143.42	56.45	10.23	136.56
Market			35.76	306.39	89.65	40.98	12.54	

Table B.2 Results as at 2007-06-14 based on 500 000 runs and $T = 10$.

B Supplementary CDO pricing results

2007-06-19	Dependence		CDO upfront and spreads $\hat{s}_{5,j}^{CDO,f}$					Error
Copula	$\hat{\vartheta}$	ρ in %	$j=1$	$j=2$	$j=3$	$j=4$	$j=5$	D_2
A	—	—	—	—	—	—	—	—
	— —	— —	—	—	—	—	—	—
$\hat{A}$	0.69	3.47	7.11	113.63	1.79	0.00	0.00	84.80
	0.10 0.94	0.19 17.16	7.16	98.78	9.47	0.62	0.01	61.64
C	1.86	3.12	7.16	91.73	16.69	3.02	0.19	53.48
	1.77 1.99	2.99 3.34	7.11	88.63	15.36	2.58	0.17	49.51
$\hat{C}$	0.08	3.08	7.12	84.37	18.35	4.62	0.54	45.84
	0.08 0.08	2.96 3.22	7.11	83.17	17.77	4.55	0.52	44.15
opC	1.06	7.52	7.10	34.03	15.94	10.55	6.18	25.67
	1.03 1.09	4.40 11.40	7.16	47.34	20.11	10.56	4.07	14.56
$\hat{opC}$	—	—	—	—	—	—	—	—
	— —	— —	—	—	—	—	—	—
F	2.68	3.10	7.13	86.91	19.93	5.08	0.50	49.54
	2.63 2.78	3.02 3.23	7.16	84.20	18.56	4.65	0.43	45.96
G	1.06	7.64	7.10	33.11	16.44	11.13	6.60	28.10
	1.03 1.10	4.20 11.89	7.13	47.09	20.78	10.86	4.17	15.76
$\hat{G}$	1.24	3.09	7.15	101.16	10.87	0.64	0.01	62.61
	1.23 1.27	2.90 3.59	7.15	97.15	10.21	0.59	0.01	59.30
J	1.06	7.84	7.12	30.03	15.93	10.73	6.32	30.00
	1.03 1.11	3.82 12.63	7.12	47.25	21.29	10.90	4.01	15.98
$\hat{J}$	—	—	—	—	—	—	—	—
	— —	— —	—	—	—	—	—	—
Ga	0.19	3.09	7.16	83.56	18.97	4.88	0.59	45.34
	0.19 0.20	3.03 3.28	7.15	82.38	18.41	5.08	0.68	43.30
Market			7.12	47.28	12.34	5.77	2.12	

Table B.3 Results as at 2007-06-19 based on 500 000 runs and $T = 5$.

2007-06-19	Dependence		CDO upfront and spreads $\hat{s}_{10,j}^{\text{CDO},f}$					Error
Copula	$\hat{\vartheta}$	ρ in %	$j=1$	$j=2$	$j=3$	$j=4$	$j=5$	D_2
A	0.74	4.86	35.50	357.50	147.14	61.35	11.44	128.06
	0.70 0.86	4.64 5.69	35.51	356.62	145.40	59.90	10.97	124.44
Â	0.29	2.52	35.51	521.71	103.91	3.72	0.01	279.19
	0.02 0.84	0.11 20.25	35.46	414.86	145.16	40.43	2.95	173.93
C	0.66	4.48	35.46	375.57	147.21	55.63	8.44	143.48
	0.65 0.69	4.39 4.68	35.50	374.35	146.48	55.04	8.29	141.08
Ĉ	0.08	4.23	35.49	377.90	142.84	53.57	8.75	139.06
	0.07 0.08	4.15 4.37	35.50	376.04	141.16	52.91	8.63	134.98
opC	1.11	12.94	35.45	265.32	86.45	46.74	24.30	62.00
	1.05 1.17	6.42 18.93	35.50	303.50	113.76	57.01	18.53	47.38
opĈ	–	–	–	–	–	–	–	–
	– –	– –	–	–	–	–	–	–
F	1.61	6.18	35.48	316.67	139.47	69.03	19.53	92.51
	1.59 1.63	6.08 6.26	35.51	316.25	138.39	68.18	19.13	89.76
G	1.13	14.59	35.46	235.87	83.45	49.98	27.42	100.80
	1.06 1.19	7.82 20.58	35.50	278.16	108.17	58.03	21.09	70.72
Ĝ	1.11	3.11	35.46	451.83	143.76	22.93	0.42	229.53
	1.09 1.20	2.71 5.93	35.49	436.64	144.18	29.72	1.10	207.30
J	1.17	16.93	35.44	201.85	75.19	50.72	30.81	147.20
	1.02 1.30	2.67 26.20	35.51	302.37	131.15	69.62	16.70	76.68
Ĵ	–	–	–	–	–	–	–	–
	– –	– –	–	–	–	–	–	–
Ga	0.14	4.54	35.49	369.38	143.42	56.45	10.20	132.56
	0.14 0.16	4.42 5.13	35.44	367.83	142.16	56.20	10.28	129.42
Market			35.48	306.89	90.56	41.89	12.85	

Table B.4 Results as at 2007-06-19 based on 500 000 runs and $T = 10$.

B Supplementary CDO pricing results

2007-06-21	Dependence		CDO upfront and spreads $\hat{s}_{5,j}^{CDO,f}$					Error
Copula	$\hat{\vartheta}$	ρ in %	$j=1$	$j=2$	$j=3$	$j=4$	$j=5$	D_2
A	–	–	–	–	–	–	–	–
	– –	– –	–	–	–	–	–	–
$\hat{A}$	0.69	3.73	8.45	146.36	3.33	0.01	0.00	114.38
	0.06 0.95	0.12 19.47	8.52	125.19	14.81	1.33	0.03	83.42
C	2.00	3.62	8.51	113.51	24.57	5.34	0.41	77.12
	1.94 2.04	3.52 3.70	8.51	111.33	23.50	5.10	0.39	74.13
$\hat{C}$	0.09	3.59	8.49	104.90	26.06	7.49	1.08	69.95
	0.09 0.09	3.53 3.67	8.46	103.50	25.41	7.38	1.03	67.82
opC	1.07	8.98	8.50	43.97	20.83	13.79	8.00	27.30
	1.06 1.09	7.39 11.09	8.46	50.49	22.44	13.50	6.77	20.90
op$\hat{C}$	–	–	–	–	–	–	–	–
	– –	– –	–	–	–	–	–	–
F	2.90	3.66	8.51	106.43	28.01	8.08	0.97	74.12
	2.81 3.03	3.53 3.85	8.49	103.51	26.60	7.57	0.90	69.35
G	1.07	9.08	8.50	42.01	20.94	14.11	8.44	30.12
	1.05 1.10	6.99 11.83	8.45	51.11	23.68	13.75	6.51	22.75
$\hat{G}$	1.25	3.48	8.50	126.56	17.65	1.45	0.02	87.53
	1.25 1.25	3.44 3.51	8.46	125.40	17.09	1.41	0.02	85.85
J	1.08	9.38	8.47	37.87	20.40	14.03	8.43	33.64
	1.05 1.11	6.65 12.70	8.51	50.58	24.47	14.01	6.46	23.22
$\hat{J}$	–	–	–	–	–	–	–	–
	– –	– –	–	–	–	–	–	–
Ga	0.21	3.68	8.46	103.13	26.82	8.24	1.27	69.49
	0.20 0.25	3.39 4.62	8.50	102.14	25.98	7.87	1.16	67.39
Market			8.48	50.47	13.28	6.11	2.43	

Table B.5 Results as at 2007-06-21 based on 500 000 runs and $T=5$.

2007-06-21	Dependence		CDO upfront and spreads $\hat{s}_{10,j}^{\text{CDO},f}$					Error
Copula	$\hat{\vartheta}$	ρ in %	$j=1$	$j=2$	$j=3$	$j=4$	$j=5$	D_2
A	0.74	5.17	37.74	387.17	166.10	72.33	14.70	154.53
	0.71 0.84	4.99 5.88	37.78	387.57	164.71	71.41	14.31	152.23
$\hat{\text{A}}$	0.28	2.54	37.81	571.95	133.91	6.37	0.02	331.42
	0.01 0.84	0.11 20.92	37.77	452.67	169.02	51.23	4.24	211.06
C	0.66	4.73	37.79	409.77	168.29	67.28	11.16	176.56
	0.65 0.67	4.67 4.79	37.77	408.06	166.62	65.95	10.70	172.31
$\hat{\text{C}}$	0.08	4.39	37.82	413.65	162.55	63.77	10.95	171.39
	0.08 0.08	4.36 4.43	37.74	411.28	161.56	63.10	10.84	167.48
opC	1.11	13.39	37.78	296.41	99.07	52.82	27.16	55.79
	1.06 1.16	8.32 17.97	37.80	326.91	121.30	61.71	22.92	52.17
op$\hat{\text{C}}$	–	–	–	–	–	–	–	–
	– –	– –	–	–	–	–	–	–
F	1.64	6.63	37.82	344.95	156.37	80.10	24.21	119.87
	1.61 1.65	6.52 6.72	37.79	343.37	154.79	78.67	23.39	114.46
G	1.13	15.13	37.78	265.07	94.93	56.74	31.02	95.82
	1.07 1.20	8.38 21.09	37.77	307.71	121.34	66.04	24.67	77.49
$\hat{\text{G}}$	1.10	3.16	37.80	492.80	169.68	30.78	0.72	262.93
	1.09 1.19	2.78 6.01	37.81	477.24	168.91	38.32	1.59	238.19
J	1.18	17.54	37.81	227.78	83.68	55.87	33.65	146.12
	1.04 1.30	4.05 26.29	37.79	325.28	140.60	76.11	20.82	85.39
$\hat{\text{J}}$	–	–	–	–	–	–	–	–
	– –	– –	–	–	–	–	–	–
Ga	0.14	4.75	37.76	402.95	162.80	66.90	12.87	162.15
	0.14 0.17	4.50 5.91	37.81	402.18	162.04	66.37	12.72	160.24
Market			37.78	329.46	97.45	44.78	14.07	

Table B.6 Results as at 2007-06-21 based on 500 000 runs and $T=10$.

B Supplementary CDO pricing results

2007-06-26	Dependence		CDO upfront and spreads $\hat{s}_{5,j}^{CDO,f}$					Error
Copula	$\hat{\vartheta}$	ρ in %	$j=1$	$j=2$	$j=3$	$j=4$	$j=5$	D_2
A	–	–	–	–	–	–	–	–
	– –	– –	–	–	–	–	–	–
$\hat{A}$	0.62	3.05	11.85	120.43	1.64	0.00	0.00	81.88
	0.15 0.91	0.34 14.21	11.86	107.07	8.85	0.43	0.00	60.87
C	1.40	2.71	11.83	102.26	16.19	2.41	0.11	46.65
	1.34 1.44	2.60 2.78	11.88	98.91	15.35	2.23	0.10	44.32
$\hat{C}$	0.07	2.66	11.83	96.50	18.29	3.98	0.35	41.02
	0.06 0.08	2.43 3.14	11.90	92.15	17.03	3.71	0.31	35.72
opC	1.05	6.65	11.89	38.95	16.72	10.89	6.32	31.88
	1.01 1.10	2.10 12.30	11.84	63.13	24.69	11.10	2.73	13.20
op$\hat{C}$	–	–	–	–	–	–	–	–
	– –	– –	–	–	–	–	–	–
F	2.09	2.63	11.85	93.76	19.32	4.44	0.39	38.81
	2.04 2.14	2.55 2.69	11.90	91.12	18.39	4.21	0.36	35.50
G	1.05	6.80	11.88	37.07	17.36	11.53	6.77	35.51
	1.01 1.11	1.35 13.52	11.83	63.41	25.78	11.24	2.34	14.52
$\hat{G}$	1.20	2.69	11.88	110.47	8.89	0.31	0.00	64.36
	1.15 1.37	2.02 5.66	11.88	105.95	10.76	0.71	0.01	57.55
J	1.06	7.16	11.85	33.84	17.53	11.98	7.11	39.69
	1.01 1.12	1.31 13.38	11.89	62.23	25.65	11.37	2.27	15.79
$\hat{J}$	–	–	–	–	–	–	–	–
	– –	– –	–	–	–	–	–	–
Ga	0.16	2.63	11.91	93.68	18.36	4.00	0.38	38.22
	0.16 0.17	2.57 2.81	11.90	92.33	17.81	4.22	0.46	36.02
Market			11.87	63.70	16.26	7.34	3.18	

Table B.7 Results as at 2007-06-26 based on 500 000 runs and $T=5$.

2007-06-26	Dependence		CDO upfront and spreads $\hat{s}_{10,j}^{\text{CDO},f}$					Error
Copula	$\hat{\vartheta}$	ρ in %	$j=1$	$j=2$	$j=3$	$j=4$	$j=5$	D_2
A	0.69	5.02	40.74	411.25	177.29	78.31	16.06	143.21
	0.68 0.71	4.99 5.20	40.73	411.39	175.83	76.82	15.54	140.92
$\hat{A}$	0.25	2.31	40.71	604.37	145.24	7.17	0.02	335.33
	0.01 0.82	0.06 19.91	40.72	481.77	181.24	54.77	4.35	205.84
C	0.60	4.49	40.69	438.91	180.49	70.93	11.21	171.53
	0.58 0.62	4.38 4.65	40.70	437.51	178.12	68.92	10.83	166.14
$\hat{C}$	0.07	4.12	40.71	444.64	175.08	67.60	11.16	168.57
	0.07 0.07	4.09 4.12	40.70	443.13	173.08	66.05	10.85	163.82
opC	1.10	12.72	40.71	329.57	107.37	55.38	28.01	53.53
	1.09 1.12	10.99 14.19	40.74	340.45	115.23	57.88	26.12	49.00
op$\hat{C}$	–	–	–	–	–	–	–	–
	– –	– –	–	–	–	–	–	–
F	1.55	6.47	40.68	366.21	164.93	84.34	25.11	99.94
	1.53 1.57	6.37 6.58	40.73	366.73	164.42	83.42	25.01	98.93
G	1.13	14.58	40.69	292.88	99.92	58.09	31.68	104.05
	1.06 1.19	7.72 20.57	40.74	338.85	130.37	69.31	24.82	75.87
$\hat{G}$	1.09	2.92	40.73	528.76	181.34	31.33	0.64	271.04
	1.08 1.19	2.45 6.14	40.73	507.97	179.39	40.29	1.66	238.31
J	1.18	17.04	40.70	255.19	87.78	57.90	34.75	156.77
	1.04 1.29	4.04 25.30	40.67	351.41	147.18	79.13	21.87	86.99
$\hat{J}$	1.30	2.30	40.69	610.02	140.20	6.55	0.01	336.57
	1.30 1.30	2.30 2.30	40.69	610.02	140.20	6.55	0.01	336.57
Ga	0.13	4.49	40.73	434.44	175.08	70.74	13.10	159.56
	0.12 0.18	4.08 6.36	40.67	433.09	173.50	69.83	12.68	156.15
Market			40.70	365.44	108.43	50.25	16.53	

Table B.8 Results as at 2007-06-26 based on 500 000 runs and $T = 10$.

B Supplementary CDO pricing results

Bibliography

Abate, J. and Valkó, P. P. (2004), "Multi-precision Laplace transform inversion", International Journal for Numerical Methods in Engineering, 60, 979–993.

Abate, J. and Whitt, W. (1992), "Numerical Inversion of Probability Generating Functions", Operations Research Letters, 12(4), 245–251.

Abate, J., Choudhury, G. L., and Whitt, W. (2006), "On the Laguerre Method for Numerically Inverting Laplace Transforms", http://www.columbia.edu/~ww2040/Laguerre1996.pdf (2009-12-30).

Abel, N. H. (1826), "Untersuchung der Functionen zweier unabhängig veränderlichen Größen x und y, wie f(x,y), welche die Eigenschaft haben, daß f(z,f(x,y)) eine symmetrische Function von z, x und y ist", Journal für die Reine und Angewandte Mathematik, 1, 11–15.

Aczél, J. (1948), "Sur les opérations définies pour nombres réels", Bulletin de la Société Mathématique de France, 76, 59–64.

Albrecher, H., Ladoucette, S., and Schoutens, W. (2007), "A Generic One-Factor Lévy Model for Pricing Synthetic CDOs", Advances in Mathematical Finance, ed. by Fu, M. C., Jarrow, R. A., Yen, J.-Y. J., and Elliott, R. J., Birkhäuser, 259–277.

Ali, M. M., Mikhail, N. N., and Haq, M. S. (1978), "A Class of Bivariate Distributions Including the Bivariate Logistic", Journal of Multivariate Analysis, 8, 405–412.

Apostol, T. M. (1974), "Mathematical Analysis", Addison Wesley.

Arendt, W., Batty, C. J. K., Hieber, M., and Neubrander, F. (2002), "Vector-Valued Laplace Transforms and Cauchy Problems", Birkhäuser.

Barbe, P., Genest, C., Ghoudi, K., and Rémillard, B. (1996), "On Kendall's Process", Journal of Multivariate Analysis, 58, 197–229.

Barndorff-Nielson, O. E. and Shephard, N. (2001), "Normal modified Stable processes", http://www.economics.ox.ac.uk/research/WP/PDF/paper072.pdf (2009-12-30).

Bauer, H. (1992), "Maß- und Integrationstheorie", 2nd ed., Gruyter.

Bernstein, S. N. (1928), "Sur les fonctions absolument monotones", Acta Mathematica, 52, 1–66.

Bibliography

Bielecki, T. and Rutkowski, M. (2002), "Credit Risk: Modeling, Valuation and Hedging", Springer.

Bondesson, L. (1982), "On simulation from infinitely divisible distributions", Advances in Applied Probability, 14, 855–869.

Borak, S., Härdle, W., and Weron, R. (2005), "Stable Distributions", http://sfb649.wiwi.hu-berlin.de/papers/pdf/SFB649DP2005-008.pdf (2009-12-30).

Brix, A. (1999), "Generalized gamma measures and shot-noise Cox processes", Advances in Applied Probability, 31, 929–953.

Cambanis, S., Huang, S., and Simons, G. (1981), "On the Theory of Elliptically Contoured Distributions", Journal of Multivariate Analysis, 11, 368–385.

Chambers, J. M., Mallows, C. L., and Stuck, B. W. (1976), "A Method for Simulating Stable Random Variables", Journal of the American Statistical Association, 71(354), 340–344.

Choe, G. H. and Jang, H. J. (2009), "Efficient algorithms for basket default swap pricing with multivariate Archimedean copulas", submitted, http://papers.ssrn.com/sol3/papers.cfm?abstract_id=1414111 (2009-12-30).

Choroś, B., Härdle, W., and Okhrin, O. (2009), "CDO Pricing with Copulae", Bulletin of the International Statistical Institute, 57, http://sfb649.wiwi.hu-berlin.de/papers/pdf/SFB649DP2009-013.pdf (2009-12-30).

Clayton, D. G. (1978), "A model for association in bivariate life tables and its application in epidemiological studies of familial tendency in chronic disease incidence", Biometrika, 65(1), 141–151.

Cohen, A. M. (2007), "Numerical Methods for Laplace Transform Inversion", Springer.

Cuadras, C. M. and Augé, J. (1981), "A continuous general multivariate distribution and its properties", Communications in Statistics—Theory and Methods, 10(4), 339–353.

Damien, P., Laud, P. W., and Smith, A. F. M. (1995), "Approximate Random Variate Generation from Infinitely Divisible Distributions with Applications to Bayesian Inference", Journal of the Royal Statistical Society: Series B (Statistical Methodology), 57(3), 547–563.

Davies, B. and Martin, B. (1979), "Numerical Inversion of the Laplace Transform: A Survey and Comparison of Methods", Journal of Computational Physics, 33(1), 1–32.

Demarta, S. and McNeil, A. J. (2004), "The t Copula and Related Copulas", http://www.math.ethz.ch/~mcneil/ftp/tCopula.pdf (2009-12-30).

Devroye, L. (1986), "Non-Uniform Random Variate Generation", Springer.

Doetsch, G. (1970), "Einführung in Theorie und Anwendung der Laplace-Transformation", Birkhäuser.

Bibliography

Donnelly, C. and Embrechts, P. (2009), "The devil is in the tails: actuarial mathematics and the subprime mortgage crisis", submitted, http://www.math.ethz.ch/~donnelly/devil09.pdf (2009-12-30).

Doob, J. L. (1990), "Stochastic Processes", Wiley-Interscience.

Duffie, D. (2007), "Innovations in Credit Risk Transfer: Implications for Financial Stability", http://www.stanford.edu/~duffie/BIS.pdf (2009-12-30).

Durrett, R. (2004), "Probability: Theory and Examples", 3rd ed., Duxbury Press.

Elstrodt, J. (2005), "Maß- und Integrationstheorie", Springer.

Embrechts, P., Klüppelberg, C., and Mikosch, T. (1997), "Modelling Extremal Events for Insurance and Finance", Springer.

Embrechts, P., Lindskog, F., and McNeil, A. J. (2001), "Modelling Dependence with Copulas and Applications to Risk Management", http://www.risklab.ch/ftp/papers/DependenceWithCopulas.pdf (2009-12-30).

Embrechts, P., McNeil, A. J., and Straumann, D. (2002), "Correlation and Dependence in Risk Management: Properties and Pitfalls", Risk Management: Value at Risk and Beyond, ed. by Dempster, M., Cambridge University Press, 176–223.

Fang, K.-T., Kotz, S., and Ng, K.-W. (1989), "Symmetric Multivariate and Related Distributions", Chapman & Hall/CRC.

Feller, W. (1971), "An Introduction to Probability Theory and Its Applications", 2nd ed., vol. 2, Wiley.

Féron, R. (1956), "Sur les tableaux de corrélation dont les marges sont données, cas de l'espace à trois dimensions", Publications de l'Institut de Statistique de l'Université de Paris, 5, 3–12.

Fisher, N. I. (1997), "Copulas", Encyclopedia of Statistical Sciences, 1, 159–163.

Flanagan, C. and Sam, T. (2002), "CDO Handbook", http://www2.wu-wien.ac.at/vgsf/curriculum/CDO%20Handbook.pdf (2009-12-30).

Folland, G. B. (1999), "Real Analysis: Modern Techniques and Their Applications", Wiley-Interscience.

Frank, M. J. (1979), "On the simultaneous associativity of $F(x,y)$ and $x + y - F(x,y)$", Aequationes Mathematicae, 19, 194–226.

Fréchet, M. (1935), "Généralisations du théorème des probabilités totales", Fundamenta Mathematica, 25, 379–387.

Fréchet, M. (1951), "Sur les tableaux de corrélations dont les marges sont données", Annales de l'Université de Lyon, 9, 53–77.

Bibliography

Gaver, D. P. (1966), "Observing stochastic processes and approximate transform inversion", Operations Research, 14, 444–459.

Genest, C. and MacKay, R. J. (1986), "Copules archimédiennes et familles de lois bidimensionnelles dont les marges sont données", The Canadian Journal of Statistics, 14, 145–159.

Genest, C. and Rivest, L.-P. (1993), "Statistical Inference Procedures for Bivariate Archimedean Copulas", Journal of the American Statistical Association, 88(423), 1034–1043.

Genest, C., Gendron, M., and Bourdeau-Brien, M. (2009), "The Advent of Copulas in Finance", The European Journal of Finance, http://www.informaworld.com/10.1080/13518470802 604457 (2009-12-30).

GSL, http://www.gnu.org/software/gsl/ (2009-12-30).

Gumbel, E. J. (1960), "Distributions des valeurs extrêmes en plusiers dimensions", Publications de l'Institut de Statistique de l'Université de Paris, 9, 171–173.

Höcht, S. and Zagst, R. (2009), "Pricing Distressed CDOs with Stochastic Recovery", http://www.mathfinance.ma.tum.de/papers/Pricing%20distressed%20CDOs%20with%20 stochastic%20recovery.pdf (2009-12-30).

Hoeffding, W. (1940), "Massstabinvariante Korrelationstheorie", Schriften des Mathematischen Seminars und des Instituts für Angewandte Mathematik der Universität Berlin, 5, 181–233.

Hofert, M. (2008), "Sampling Archimedean copulas", Computational Statistics & Data Analysis, 52(12), 5163–5174.

Hofert, M. and Scherer, M. (2010), "CDO pricing with nested Archimedean copulas", Quantitative Finance, in press.

Hougaard, P. (1986), "Survival models for heterogeneous populations derived from stable distributions", Biometrika, 73(2), 387–396.

Hull, J. C. (2008), "Options, Futures, and Other Derivatives", 7th ed., Prentice Hall Series in Finance.

Hull, J. C., Predescu, M., and White, A. (2006), "The valuation of correlation-dependent credit derivatives using a structural approach", http://www.rotman.utoronto.ca/%7Ehull/DownloadablePublications/StructuralModel.pdf (2009-12-30).

Hult, H. and Lindskog, F. (2002), "Multivariate extremes, aggregation and dependence in elliptical distributions", Advances in Applied Probability, 34, 587–608.

Joe, H. (1993), "Parametric families of multivariate distributions with given margins", Journal of Multivariate Analysis, 46, 262–282.

Joe, H. (1997), "Multivariate Models and Dependence Concepts", Chapman & Hall/CRC.

Joe, H. and Hu, T. (1996), "Multivariate Distributions from Mixtures of Max-Infinitely Divisible Distributions", Journal of Multivariate Analysis, 57, 240–265.

Kalemanova, A., Schmid, B., and Werner, R. (2007), "The Normal Inverse Gaussian Distribution for Synthetic CDO Pricing", The Journal of Derivatives, 14(3), 80–93.

Kemp, A. W. (1981), "Efficient Generation of Logarithmically Distributed Pseudo-Random Variables", Journal of the Royal Statistical Society: Series C (Applied Statistics), 30(3), 249–253.

Kiesel, R. and Scherer, M. (2007), "Dynamic Credit Portfolio Modelling in Structural Models with Jumps", http://numerik.uni-ulm.de/preprints/2008/Kiesel_Scherer_Dec07.pdf (2009-12-30).

Kimberling, C. H. (1974), "A probabilistic interpretation of complete monotonicity", Aequationes Mathematicae, 10, 152–164.

Klenke, A. (2008), "Probability Theory: A Comprehensive Course", Springer.

Kortschak, D. and Albrecher, H. (2009), "Asymptotic Results for the Sum of Dependent Non-identically Distributed Random Variables", Methodology and Computing in Applied Probability, 11, 279–306.

Kotz, S. and Nadarajah, S. (2004), "Multivariate t Distributions and Their Applications", Cambridge University Press.

Li, D. X. (2000), "On Default Correlation: A Copula Function Approach", The Journal of Fixed Income, 9(4), 43–54.

Li, H. (2008), "Tail dependence comparison of survival Marshall-Olkin copulas", Methodology and Computing in Applied Probability, 10(1), 49–54.

Li, X., Mikusiński, P., and Taylor, M. D. (2002), "Some integration-by-parts formulas involving 2-copulas", Distributions with Given Marginals and Statistical Modelling, ed. by Cuadras, C. M., Fortiana, J., and Rodríguez-Lallena, J. A., Kluwer Academic Publishers, 153–159.

Lindskog, F., McNeil, A. J., and Schmock, U. (2002), "Kendall's tau for elliptical distributions", http://www.math.kth.se/~lindskog/papers/KendallsTau.pdf (2009-12-30).

Ling, C. H. (1965), "Representation of associative functions", Publicationes Mathematicae Debrecen, 12, 189–212.

Lucas, D. J. (1995), "Default Correlation and Credit Analysis", Journal of Fixed Income, 4(4), 76–87.

Markit, http://www.markit.com/ (2009-12-30).

Marshall, A. W. and Olkin, I. (1967), "A multivariate exponential distribution", Journal of the American Statistical Association, 62, 30–44.

Marshall, A. W. and Olkin, I. (1988), "Families of Multivariate Distributions", Journal of the American Statistical Association, 403rd ser., 83, 834–841.

Matsumoto, M. and Nishimura, T. (1998), "Mersenne Twister: A 623-Dimensionally Equidistributed Uniform Pseudo-Random Number Generator", ACM Transactions on Modeling and Computer Simulation, 8(1), 3–30.

McCulloch, J. H. (2003), "The Risk-Neutral Measure and Option Pricing under Log-Stable Uncertainty", http://economics.sbs.ohio-state.edu/pdf/mcculloch/wp03-07.pdf (2009-12-30).

McCulloch, J. H. and Lee, S. H. (2007), "Estimation of the Risk Neutral Measure with the Stable Option Pricing Model", https://editorialexpress.com/cgi-bin/conference/download.cgi?db_name=sce2007&paper_id=305 (2009-12-30).

McNeil, A. J. (2008), "Sampling nested Archimedean copulas", Journal of Statistical Computation and Simulation, 6th ser., 78, 567–581.

McNeil, A. J. and Nešlehová, J. (2009), "Multivariate Archimedean copulas, d-monotone functions and l_1-norm symmetric distributions", The Annals of Statistics, 37(5b), 3059–3097.

McNeil, A. J., Frey, R., and Embrechts, P. (2005), "Quantitative Risk Management: Concepts, Techniques, and Tools", Princeton University Press.

Menger, K. (1942), "Statistical metrics", Proceedings of the National Academy of Sciences of the United States of America, 28, 535–537.

NAG, http://www.nag.co.uk/ (2009-12-30).

Nelsen, R. B. (2005), "Dependence Modeling with Archimedean Copulas", http://www.lclark.edu/~mathsci/brazil2.pdf (2009-12-30).

Nelsen, R. B. (2007), "An Introduction to Copulas", Springer.

Nešlehová, J. (2004), "Dependence of Non-Continuous Random Variables", PhD thesis, Carl von Ossietzky Universität Oldenburg, http://www.staff.uni-oldenburg.de/dietmar.pfeifer/JohannaDiss.pdf (2009-03-09).

Nešlehová, J. (2007), "On rank correlation measures for non-continuous random variables", Journal of Multivariate Analysis, 98, 544–567.

Nolan, J. P. (1997), "Numerical calculation of stable densities and distribution functions", Stochastic Models, 13(4), 759–774.

Nolan, J. P. (2009), "Stable Distributions—Models for Heavy Tailed Data", Birkhäuser, http://academic2.american.edu/~jpnolan/stable/chap1.pdf (2009-12-30).

Oberhettinger, F. and Badii, L. (1973), "Tables of Laplace Transforms", Springer.

Bibliography

Okhrin, O. (2007), "Hierarchical Archimedean copulas: Structure determination, properties, applications", PhD thesis, Europa-Universität Viadrina Frankfurt (Oder).

Olivares, P. and Seco, L. (2003), "Stable distributions: A survey on simulation and calibration methodologies", http://www.risklab.ca/Stableproject.pdf (2009-12-30).

Pacioli, http://pacioli.mathematik.uni-ulm.de/ (2009-12-30).

Post, E. L. (1930), "Generalized differentiation", Transactions of the American Mathematical Society, 32, 723–781.

Press, W. H., Teukolsky, S. A., Vetterling, W. T., and Flannery, B. P. (2007), "Numerical Recipes: The Art of Scientific Computing", 3rd ed., Cambridge University Press.

R, http://www.r-project.org/ (2009-12-30).

Rényi, A. (1959), "On measures of dependence", Acta Mathematica Hungarica, 10, 441–451.

Ridout, M. (2008), "Generating random numbers from a distribution specified by its Laplace transform", http://www.kent.ac.uk/IMS/personal/msr/webfiles/rlaptrans/SimRandom3.pdf (2009-12-30).

Rosiński, J. (2001), "Series representations of Lévy processes from the perspective of point processes", Lévy Processes—Theory and Applications, ed. by Barndorff-Nielsen, O. E., Mikosch, T., and Resnick, S. I., Birkhäuser, http://www.math.utk.edu/~rosinski/Manuscripts/seriesppF.pdf (2009-12-30).

Rosiński, J. (2007), "Tempering Stable processes", Stochastic Processes and their Applications, 117, 677–707.

Savu, C. and Trede, M. (2006), "Hierarchical Archimedean copulas", http://www.uni-konstanz.de/micfinma/conference/Files/papers/Savu_Trede.pdf (2009-12-30).

Scarsini, M. (1984), "On measures of concordance", Stochastica, 8(3), 201–218.

Schmitz, V. (2003), "Copulas and Stochastic Processes", PhD thesis, Rheinisch-Westfälische Technische Hochschule Aachen, http://darwin.bth.rwth-aachen.de/opus3/volltexte/2004/935/pdf/Schmitz_Volker.pdf (2009-12-30).

Schönbucher, P. J. (2003), "Credit Derivatives Pricing Models", Wiley.

Schönbucher, P. J. and Schubert, D. (2001), "Copula-Dependent Default Risk in Intensity Models", http://papers.ssrn.com/sol3/papers.cfm?abstract_id=301968 (2009-12-30).

Schoutens, W. (2003), "Lévy Processes in Finance: Pricing Financial Derivatives", Wiley.

Schröder, E. (1870), "Vier combinatorische Probleme", Zeitschrift fü Mathematik und Physik, 15, 361–376.

Bibliography

Schweizer, B. (1991), "Thirty years of copulas", Advances in Probability Distributions with Given Marginals, ed. by Dall'Aglio, G., Kotz, S., and Salinetti, G., Kluwer Academic Publishers, 13–50.

Schweizer, B. and Sklar, A. (1961), "Associative functions and statistical triangle inequalities", Publicationes Mathematicae Debrecen, 8, 169–186.

Schweizer, B. and Sklar, A. (1963), "Associative functions and abstract semigroups", Publicationes Mathematicae Debrecen, 10, 69–81.

Schweizer, B. and Sklar, A. (1983), "Probabilistic Metric Spaces", North-Holland, New York.

Schweizer, B. and Wolff, E. F. (1981), "On nonparametric measures of dependence for random variables", The Annals of Statistics, 9, 879–885.

Sklar, A. (1959), "Fonctions de répartition à n dimensions et leurs marges", Publications de L'Institut de Statistique de L'Université de Paris, 8, 229–231.

Sklar, A. (1973), "Random variables, joint distribution functions, and copulas", Kybernetika, 9, 449–460.

Sklar, A. (1996), "Random variables, distribution functions, and copulas—a personal look backward and forward", Distributions with Fixed Marginals and Related Topics, 28, 1–14.

Stehfest, H. (1970), "Algorithm 368: numerical inversion of Laplace transforms", Communications of the ACM, 13(1), 47–49.

Szegö, G. (1975), "Orthogonal Polynomials", 4th ed., American Mathematical Society.

Talbot, A. (1979), "The accurate numerical inversion of Laplace transforms", Journal of the Institute of Mathematics and Its Applications, 23, 97–120.

Thiele, T. N. (1909), "Interpolationsrechnung", Teubner.

Tricomi, F. (1935), "Transformazione di Laplace e Polinami di Laguerre", Accademia Nazionale dei Lincei, 21, 232–239.

Tweedie, M. C. K. (1984), "An index which distinguishes between some important exponential families", Statistics: Applications and New Directions: Proceedings Indian Statistical Institute Golden Jubilee International Conference, 579–604.

UZWR, http://www.informatik.uni-ulm.de/uzwr/ (2009-12-30).

Valkó, P. P. and Abate, J. (2004), "Comparison of Sequence Accelerators for the Gaver Method of Numerical Laplace Transform Inversion", Computers and Mathematics with Applications, 48, 629–636.

Valkó, P. P. and Vojta, B. L. (2001), "The List", http://www.pe.tamu.edu/valko/public_html/NIL/LapLit.pdf (2009-12-30).

Wagner, R. (2003), "Mersenne Twister Random number Generator", http://www-personal.umich.edu/~wagnerr/MersenneTwister.html (2009-12-30).

Whelan, N. (2004), "Sampling from Archimedean Copulas", Quantitative Finance, 4(3), 339–352.

Widder, D. V. (1935), "An Application of Laguerre Polynomials", Duke Mathematical Journal, 1(2), 126–136.

Widder, D. V. (1946), "The Laplace Transform", Princeton University Press.

Williams, D. (1991), "Probability with Martingales", Cambridge University Press.

Wimp, J. (1981), "Sequence transformations and their applications", Academic Press.

Wu, F., Valdez, E. A., and Sherris, M. (2006), "Simulating Exchangeable Multivariate Archimedean Copulas and its Applications", Communications in Statistics—Simulation and Computation, 36(5), 1019–1034.

Wynn, P. (1956), "On a Procrustean technique for the numerical transformation of slowly convergent sequences and series", Mathematical Proceedings of the Cambridge Philosophical Society, 52(4), 663–671.

Bibliography

Index

Symbols

$\bar{\mathbb{R}}$ 16
$\Delta_{(a,b)} H$ 16
$\mathrm{D}_{j_k \ldots j_1} H$ 18
I 19
$\bar{H}$ 19
T^- 19
ran 19
W 24
M 24
Π 26
$\hat{C}$ 31
$C^{(i_1 \ldots i_j)}$ 31
$\preceq$ 35
ρ_S 36
τ 37
sgn 37
$\hat{\tau}$ 37
λ_l 39
λ_u 39
ψ 48
Ψ 48
Ψ_∞ 54
F_0 57
ψ_{01} 57
F_{0s} 58
F_{01} 58
A 61
C 61
F 61
G 61
J 61
$\mathrm{Geo}(p)$ 62
$\mathrm{Log}(p)$ 62
$\Gamma(\alpha, \beta)$ 62
$\mathrm{S}(\alpha, \beta, \gamma, \delta; 1)$ 63
$x_{\min}$ 82
κ 82
$x_{\max}$ 83
MXRE 84
MRE 84
$\tilde{\mathrm{S}}(\alpha, \beta, \gamma, \delta, h; 1)$ 107
opC 126
Ga 126
t_4 126
λ_i 137
p_i 137
τ_i 137
ρ_{kl} 139
ρ_s 140
I 141
T 141

Index

$\mathcal{T}$ 141
R 141
L_t 141
$L_{t,j}$ 141
l_j 141
u_j 141
N_t 141
$N_{t,j}$ 141
s_T^{pCDS} 141
EDPL_T 142
EDDL_T 142
$\Delta_{(t_{k-1}, t_k]}$ 142
$\text{EDPL}_{T,j}$ 142
$\text{EDDL}_{T,j}$ 142
$s_{T,j}^{\text{CDO}}$ 142
$s_T^{\text{pCDS},f}$ 142
$s_{T,j}^{\text{CDO},f}$ 142
$\overline{\text{EDPL}}_{T,j}$ 144
$\overline{\text{EDDL}}_{T,j}$ 144
$\hat{s}_{T,j}^{\text{CDO},f}$ 144
ε_{CDO} 145
$s_{T,j}^{\text{CDO},m}$ 145
D_1 145
D_2 145
$s_T^{\text{pCDS},m}$ 146
$\mathbb{K}$ 160
$V_{[a,b]}$ 160
BV 160
BV_{loc} 162
$\mathcal{LS}$ 162
abs 162
L^1_{loc} 163
$\mathcal{L}$ 163
$\mathcal{L}^{-1}$ 164
$\mathcal{LS}^{-1}$ 164

A

abscissa of convergence 162, 163
absolutely continuous 161
absolutely continuous component 25
absolutely continuous copulas 26
absolutely convergent 164
absolutely monotone 50
accelerative 75
accrued interest 141
Archimedean copulas 48
Archimedean generator 48
Archimedean property 50
Archimedean t-norm 50
associative 49
attachment points 141

B

basis points 135
Bernstein's Theorem 53
beta function 102
bounded variation 160
Bromwich integral 164

C

CDO market spreads 145
CDO tranche spreads 142
CDS 137
characteristic generator 42
collateralized debt obligation 135
comonotone 30
completely monotone 50
concordance ordering 36
concordant 37
conditional distribution function 158

Index

conditional distribution functions 40
conditional distribution method 40
conditional expectation 157, 158
conditional probability 157, 158
copula 22
copula of 28
correlation 33
correlation coefficient 33
countermonotone 30
credit default swaps 137
Cuadras-Augé copula 159

D

d-increasing 17, 22
d-monotone 50
Debye function of order one 64
default correlation 139
default probability 137
default times 137
defective 165
delta-convergent sequence of densities
.... 74
diagonal copula 27
disconcordant 37
discount factors 141
dispersion matrix 42
distribution function 19

E

elliptical copulas 44
elliptical distributions 42
equity tranche 135
exchangeable 32
expected discounted default leg 142

expected discounted default leg of a
    CDO tranche 142
expected discounted premium leg 142
expected discounted premium leg of a CDO
    tranche 142
explicit copulas 44
exponential integral 64
exponentially tilted distribution 105
exponentially tilted stable distributions
.... 107
exponentially tilting 77
extreme-value copulas 64

F

fair spread 142
fast rejection algorithm 106
Fixed Talbot algorithm 79
Fréchet bounds 24
Frequency Differentiation Theorem ... 162
fully-nested Archimedean copula 55

G

gamma distribution 62
Gaussian copula 44
Gaver functionals 74
Gaver-Stehfest algorithm 75, 79
Gaver-Wynn-rho algorithm 76, 80
generalized inverse 19, 43
generator 48
geometric distribution 62
grounded 16, 21

H

H-volume 16
hierarchical Archimedean copulas 57
hierarchies 55

189

Index

I

i.i.d. 68
ill-posed 72
increasing function 16
independence copula 28
inner distribution functions 58
inner generators 58
inner power families 66
Integration and Differentiation Theorem 163
Integration by Parts Formula 161
intensity 137
inversion method 21

K

Kendall's tau 37
Kendall's transformation 59

L

Laguerre coefficients 77
Laguerre functions 77
Laguerre series 77
Laguerre-series algorithm 78, 81
Laplace transform 163
Laplace transformable 163
Laplace-Stieltjes transform 162
Laplace-Stieltjes transformable 162
Lattice-Poisson algorithm 77
Lebesgue-Stieltjes integrable 160
Lebesgue-Stieltjes integral 160
Lebesgue-Stieltjes measure 160
less concordant 36
less positive lower orthant dependent .. 35
location vector 42
logarithmic convergence 75
logarithmic distribution 62

loss affecting a CDO tranche 141
lower attachment points 141
lower Fréchet bound 24
lower tail dependent 39
lower tail independent 39
lower tail-dependence coefficient 39

M

margin 17
Marshall-Olkin algorithm 68
Marshall-Olkin copulas 159
maximal relative error 84
McNeil's algorithm 69
mean relative error 84
measure of concordance 36
measure of dependence 34
measures of association 32
more concordant 36
more positive lower orthant dependent
.... 35

N

nested Archimedean copulas 55
nested outer power Archimedean copulas 89
nesting condition 57
nesting levels 55
non-degenerate 42

O

outer distribution function 58
outer generator 58
Outer power copulas 66
outer power families 66

P

partially-nested Archimedean copula ... 55

payment schedule 141
perfect negative dependence 30
perfect positive dependence 30
perfectly linearly dependent 33
portfolio CDS market spread 146
portfolio CDSs 137
portfolio-CDS spread 141
portfolio-loss process 141
positive lower orthant dependent 35
Post's Formula 165
probability distribution function 19
probability integral transformation 59
protection buyer 135
protection seller 135
pseudo-inverse 48

Q
quantile function 19

R
radially symmetric 31
random sample 37
rank correlation measures 34
ranks 34
rational copulas 64
reciprocal differences 76
recovery rate 141
regular conditional distribution 158
rejection algorithm 105
remaining nominal 141
remaining nominal of a CDO tranche
 141
Riemann-Stieltjes integrable 161

S
Salzer weights 75

sample versions 37
Schröder's fourth problem 59
sectors 56
securitization 135
serial iterates 50
singular component 26
singular copulas 26
Sklar's Theorem 26
Sklar's Theorem for survival functions
 31
Spearman's rho 36
spread 135
stable distribution 62
standard extension copula 22
Standard rejection 105
step function 161
strict 48, 49
subadditive 59
subcopula 21
survival copulas 31
survival functions 19
survival probability 137
synthetic CDOs 137

T
t_ν copula 45
t-norm 49
table look-up methods 83
tail dependence 39
tail-dependence coefficients 39
Thiele's interpolation formula 76
tilted Archimedean generator 104
time to maturity 141
tranches 135
triangular norm 49

Index

trigger variables 138
truncation 82

U

uniform margins 21
Uniqueness Theorem 164
upfront payment 142
upper attachment points 141
upper Fréchet bound 24
upper incomplete gamma function 64
upper tail dependent 39
upper tail independent 39
upper tail-dependence coefficient 39

V

version 157

Die VDM Verlagsservicegesellschaft sucht für wissenschaftliche Verlage abgeschlossene und herausragende

Dissertationen, Habilitationen, Diplomarbeiten, Master Theses, Magisterarbeiten usw.

für die kostenlose Publikation als Fachbuch.

Sie verfügen über eine Arbeit, die hohen inhaltlichen und formalen Ansprüchen genügt, und haben Interesse an einer honorarvergüteten Publikation?

Dann senden Sie bitte erste Informationen über sich und Ihre Arbeit per Email an *info@vdm-vsg.de*.

Sie erhalten kurzfristig unser Feedback!

VDM Verlagsservicegesellschaft mbH
Dudweiler Landstr. 99   Telefon +49 681 3720 174
D - 66123 Saarbrücken   Fax   +49 681 3720 1749
www.vdm-vsg.de

Die VDM Verlagsservicegesellschaft mbH vertritt

Printed by Books on Demand GmbH, Norderstedt / Germany